THE INVISIBLE
BUREAUCRACY

THE INVISIBLE BUREAUCRACY

The Unconscious in Organizational Problem Solving

Howell S. Baum, Ph.D.
University of Maryland, Baltimore

New York   Oxford
OXFORD UNIVERSITY PRESS
1987

Oxford University Press

Oxford   New York   Toronto
Delhi   Bombay   Calcutta   Madras   Karachi
Petaling Jaya   Singapore   Hong Kong   Tokyo
Nairobi   Dar es Salaam   Cape Town
Melbourne   Auckland

and associated companies in
Beirut   Berlin   Ibadan   Nicosia

Library of Congress Cataloging-in-Publication Data

Baum, Howell S.
   The invisible bureaucracy.

   Bibliography: p.
   Includes index.
   1. Organization.   2. Problem solving.   3. Bureauc-
racy—Psychological aspects.   I. Title.
HM131.B375   1987   302.3'5   86-33133
ISBN 0-19-503961-0

Parts of Chapter 4 appeared previously in "Autonomy, Shame and Doubt: Power in the Bureaucratic Lives of Planners," *Administration and Society,* 15, no. 2 (August 1983), pp. 147–84, and are reprinted here with the permission of Sage Publications Inc.

Parts of Chapter 5 appeared previously in "Predicaments of Bureaucratic Subordinacy: Dramatic Escape and Compensation," *Political Psychology,* 6, no. 4 (December 1985), pp. 681–703, and are reprinted here with the permission of the International Society of Political Psychology.

Parts of Chapter 7 appeared previously in "The Advisor as Invited Intruder," *Public Administration Review,* 42, no. 6 (November/December 1982), pp. 546–52, and are reprinted with permission from Public Administration Review. Copyright © 1982 by the American Society for Public Administration, Washington, D.C. All rights reserved.

Quotations from James George Frazer, *The Golden Bough,* abridged edition (New York: Macmillan, 1940), are reprinted with the permission of the Macmillan Publishing Company and A. P. Watt, Ltd.

9 8 7 6 5 4 3 2 1

Printed in the United States of America
on acid-free paper

To Madelyn and Elena

Preface

This book grew from an earlier study of planners. I was originally interested in understanding how planners conceptualize their work and make sense of their environment. To my surprise, I found that many planners do not regard their work as having a social or organizational environment. Rather, they concentrate on analyzing data in a world of pure information. When I attempted to explain this finding in an earlier book (Baum, 1983a), I focused on planners' abilities as practitioners. Some people who choose to become planners may be singularly unsuited for the work because they react anxiously to the exercise of power—which is really the essence of the planning environment—and retreat to the security of narrow technical work.

Still, despite the apparent validity of my analysis, I felt that it tended excessively to blame planners for difficulties in their work. I wondered whether there were not something about the bureaucratic setting itself which makes it problematic for most people to think about power, politics, and organization. I reexamined the planners' statements and interviewed a number of public administrators with this question in mind. In particular, I sought to understand how bureaucratic workers think and feel about organizational power: what it is, who exercises it, how it is asserted, and how it may be controlled. What people said, as the book reveals, indicates that, indeed, bureaucratic organizations have an intricate psychological structure which tends to work at cross-purposes with practitioners' declared intentions of planning and solving problems rationally.

Taken together, these two books pose an important question concerning the relative influences of individual personality and organizational setting in shaping organizational performance. The books call for further study to clarify the conditions under which each of these elements matters especially. While few of us may be happy with any interpretation which suggests that we are ourselves the cause of many of our frustrations, even this book, with its focus on the bureaucratic environment, shows that organizational settings often constrain us only because we collude in accepting certain limits. Although this book shows the unconscious world to be complex almost beyond our most careful comprehension, the lesson is that, if we want to work more effectively, we must take more responsibility, for as much as we are willing to understand.

Many friends and colleagues have helped challenge and clarify the ideas which have emerged in the book. John Forester, whose interests in planning and organizing are long-standing, both provided sophisticated substantive comments on the manuscript and offered sensible suggestions to improve its readability. Beryl Radin drew on her familiarity with administrative literature and her years of organizational experience to raise good questions about assumptions I took for granted. Michael Diamond, a partner in the effort to understand organizations psychoanalytically, provided enthusiastic support, reviewed the manuscript, and offered helpful suggestions. Leonard Duhl, who originally got me to think about psychological issues and organizational politics, accepted their connections without question but pushed me to develop more stories and illustrations to demonstrate the connections to others. Donald Levine reviewed the manuscript and helpfully suggested clarifying the audience for the book and choosing an appropriate language for presenting cross-disciplinary material.

Many others offered useful comments on parts of the manuscript: Sy Adler, Seth Allcorn, Chris Argyris, Richard Bolan, Bruce Buchanan, Larry Hirschhorn, Ralph Hummel, Gideon Kunda, Harry Levinson, Seymour Mandelbaum, Susan Schneider, Donald Schön, Howard Schwartz, Michael Teitz, and William Torbert. Two anonymous reviewers for Oxford University Press immeasurably improved the manuscript by insisting that psychoanalytic terms be presented more clearly, that unconscious phenomena be more extensively illustrated, and that, in general, the argument be presented explicitly enough to satisfy a diverse reading audience.

Ruth Young and Kenneth Corey supported this work institutionally. Gwen Young and Pat Leedham provided patient, quick, and courteous assistance in typing and revising the manuscript.

I dedicate the book to Madelyn, my wife, who has aided me in my psychoanalytic learning and, more important still, faithfully encouraged my work. I dedicate the book also to Elena, my daughter, who has shown me how complex all of us are from the very beginning and who has given me the hope that sensitive responses to needs early in life provide the strength to act effectively later on.

Baltimore   H.S.B.
April 1987

Contents

THE INVISIBLE
BUREAUCRACY

Introduction

Philadelphia—Even from the modestly appointed office he occupies here as head of the Hospital of the University of Pennsylvania, Chuck Buck can see that bear in the Catoctin Mountain Zoo of Western Maryland.

This is no ordinary bear, you see. This is Wendy, the Malayan Sun Bear, who has come to symbolize so much to Dr. Charles R. Buck, Jr., about the more than four years he spent inside a government bureaucracy as Maryland's Secretary of Health and Mental Hygiene before leaving last month.

In 1980 Wendy bit her keeper on the hand. State health regulations said that the animal should be destroyed to determine whether it has rabies. But the keeper loved the bear and offered to take rabies shots as a precaution so the animal would not have to be killed.

That arrangement was fine with Wendy and her keeper, but not so fine with health officials.

"All the way through the health department, all the way to my office, no one at various levels of authority would say 'Don't kill the bear,'" Dr. Buck recalled recently. "It took [my deputy secretary], who was a physician and I guess had less of the bad experiences with regulations that bureaucrats can have, to finally say, 'Don't kill the bear.'"

To Chuck Buck, that story illustrates a basic truth about government: Bureaucrats frequently rely on rules and regulations rather than common sense and their own judgment in making decisions. That means that governmental actions are uncompromising or unrealistic.

Why does government run by rules rather than reason? For one thing, Dr. Buck says, bureaucrats are so conditioned to make decisions by the book that even when they do have latitude, they tend not to use it. ("Public gets bureaucracy it wants, Dr. Buck says," *Baltimore Sun*, August 15, 1983)

More and more people work in bureaucratic organizations. Yet "bureaucracy" is commonly an epithet for what is wrong with jobs. People often assert that they do their work despite obstacles the organization throws up against them. Some bureaucrats are effective and even creative. Many others, however, feel that they are neither and blame the system they work in for hampering them. This book investigates that feeling and the complaints it spawns.

The book examines the subjective experience of bureaucracy, the "psychological structure" that invisibly but powerfully accompanies its social structure. The approach here is psychoanalytic, investigating what people say about work and organizations and probing for unconscious meanings of bureaucracy and accompanying feelings about it. Psychoanalysis assumes that people have both conscious and unconscious intentions, that the latter frequently conflict with the former, and that conscious aims are often frustrated by divergent unconscious goals.

In this framework, workers' relations with bureaucracy are more complicated than their complaints usually suggest. How they respond to the social structures of bureaucracy depends on both their conscious and unconscious thoughts about such matters as authority, responsibility, and regulation. Sometimes people react similarly to a given bureaucratic social structure, suggesting that certain structures produce a typical psychological response. Other reactions are primarily idiosyncratic expressions of past personal histories, although common past experiences may also lead to similar reactions to bureaucratic organizations. Insofar as certain social features of bureaucracy normally evoke specific unconscious feelings, one may conclude that bureaucracy is the immediate cause of those feelings and any resulting work frustrations. At the same time, variations in responses raise questions about whether the constraints bureaucrats feel forced to work under are actually a result of the organization or of their own private assumptions about it. In short, reactions to bureaucracy are interactions: an organization contributes social structures to which members give their interpretations.[1]

Weber's (1967) classic definition of bureaucracy describes its typical social characteristics. Work is divided into parts and allocated among relatively specialized workers. While responsibility is thus dispersed, authority is centralized, so that a relatively small number of administrators hierarchically manage the work of a relatively large number of subordinates. Because the chain of command may be quite lengthy in a large organization, codified, impersonal rules replace close personal supervision as the regulator of the quality of effort. Thus members may have limited contact with other specialists in related work, as well as with supervisors who decide on evaluation, pay, and promotion.

Organizations vary in product and technology, size, degree of specialization, allocation of authority, hierarchical levels, extent and specificity of regulations, evaluation procedures, and managerial style. At one end of a continuum are those organizations closely matching Weber's definition of a typical bureaucracy. At the other end one may find organizations where workers tend to be autonomous generalists, authority is relatively evenly allocated, the hierarchy is shallow, and decision making is more or less democratic. Probably these organizations are small, and they are more likely to deliver services than to make goods. They deviate too much from Weber's definition to be considered bureaucracies, although they may have some bureaucratic elements. In this work, "bureaucracy" will refer to any bureaucratic elements, without implying that an organization consists exclusively of them. For convenience, members of organizations with these elements will be called "bureaucrats," "bureaucratic workers," or "bureaucratic members."

Two questions about workers' interpretations of and responses to bureaucratic social structures guide this study: How do workers feel about being held responsible for tasks which they lack the necessary authority to complete? How do workers feel about being accountable for their performance to administrators whom they rarely see and about whom they know little?

Material for answers to these questions comes from planners and public administrators interviewed about their organizations and their work. Their observations indicate that typical patterns of unconscious feelings, interests, and intentions strongly affect the ways in which bureaucrats interpret their assignments, set priorities, and allocate their time and energy. Consciously, workers say that they want to develop rational strategies for solving organizational problems. However, unconsciously, workers are frequently more concerned about avoiding anxiety associated with personal relationships with others, even when doing so leads them away from their formal responsibilities. Hierarchical authority relationships make bureaucratic workers especially anxious about being shamed by superiors for poor performance. In efforts to defend themselves against shame, workers may avoid taking initiative, attempt to confuse others about their responsibilities, ritualistically recite rules, and even doubt their own competence to act. Because practitioners do these things unconsciously, generally they are unaware of their own "hidden agenda" and have little opportunity to control it, even when it leads them to act in conflict with their conscious intentions.

Many professionals choose to ignore the organizational and interpersonal environment that mixes these feelings into their work. These practitioners tend to assume that solving problems is primarily an abstract intellectual process. Although this belief may help them avoid thinking about the uncomfortable ambiguities and risks of organizational politics and power, it is self-defeating. Ignoring unconscious anxieties does not drive them away, it only keeps them underground, where they form thickets that trap the unprepared.

The costs of neglecting the psychological domain of organizations are unacceptable. These costs are understated by saying simply that planning may fail and organizational problems may not be solved. Bureaucrats may not even be seriously concerned with organizational problems and goals, nor even with their own self-interests. Instead, they may spend a lot of time unconsciously worrying about threats from others. While some of these concerns may be real, many are imaginary, the result of unrecognized unconscious assumptions about people with authority, peers, and so forth. Bureaucrats may spend a great deal of energy avoiding or managing interpersonal conflicts that are only vaguely defined, are rarely resolved, and ultimately seem neither important nor necessary. Thus workers do a lot that doesn't satisfy them, and they waste immeasurable organizational resources in the process. In the end, organizations' clients, constituents, and customers are poorly served.

In order to improve organizational productivity and the quality of bureaucratic work, it is first necessary to make this psychological domain visible, so that workers can recognize the unconscious thoughts that weigh so heavily on their actions. The more people understand what is going on around them, the more they can make reasonable choices and effective decisions about how and with whom to work.

Frequently, workers come to regard bureaucracy simply as "crazy," or incomprehensible, and they often cannot understand why they become deeply invested in activities that have little apparent justification. The book provides a language for interpreting day-to-day organizational experiences; centrally, it insists that workers' inner experiences of the organization matter. The study data pertain to planners and public administrators. However, policy analysts, systems analysts, management scientists, private sector managers, social workers, and organizational consultants bear some of the same responsibilities and encounter similar organizational structures, and this work will help make sense of their bureaucratic predicaments and responses to them. Above all, this examination should help them to understand why problem solving seems unnecessarily complicated.

The book should also assist educators in professional schools. Professional education tends to emphasize the substance of problems and formal analytic methods for solving them, while organizational and political contexts receive relatively little attention. New graduates often imagine that they will move into a world populated by information rather than people. For many the main "reality shock" is simply discovering the organizational setting.

In addition, this book should interest organizational sociologists and psychologists. Most formal organizational theory emphasizes the social structure of bureaucracy and tends to overlook its psychological dimensions. Often psychologically motivated actions are forced into sociological interpretations, or observations about these actions may be presented in afterthought, interesting but only exceptions. This book shares the sociologists' view that organizational structures consistently shape participants' actions; however, the data show that some of the most powerful structures operate psychologically, through unconscious thought and feeling. Moreover, when different participants react differently to the same social structure, they may do so because they experience different psychological structures and/or bring different personality orientations to them.

Thus this book adds to the theoretical understanding of organizational behavior in two ways. First, it presents data that are not simply sociological and that require a psychological framework for proper interpretation. Second, it brings a psychological sensitivity to traditional organizational data to show how they may be much more richly understood when viewed from two different vantage points. The psychoanalytic interpretations of cases do not reveal everything about workers' behavior, but they show a lot which would otherwise go unnoticed and much which helps make sense of commonplace but otherwise puzzling or seemingly contradictory actions. Because the data come from brief, even though intensive, interviews, some interpretation is necessarily speculative. The reader should consider the interpretation as a possible explanation for actions that must somehow be understood. Further study will help clarify the interrelationships in organizational activity.

THE PSYCHOANALYTIC PERSPECTIVE

Four specific assumptions will be particularly important in the analysis of data in the following chapters. The first is that people think both consciously and unconsciously. Substantively, people tend to think consciously primarily about matters that are easy, comfortable, or pleasant. Anything that is somehow unpleasant—perhaps frightening or embarrassing—they push out of awareness and continue to think about covertly, unconsciously. People cannot consciously control their unconscious thoughts, with the result that these thoughts may subvert espoused values and intentions. Thus people may simultaneously think in different ways about the same subjects.

For example, someone may consciously think about what he or she likes about someone else while unconsciously mulling over things he or she dislikes. This person has a conflict regarding what to think about and how to act toward the other person. One may say that this person is ambivalent toward the other, wanting, perhaps, both to approach and to avoid or to join and to attack the other. So long as someone feels ambivalent, he or she may have difficulty choosing definite actions. In particular, because certain feelings are unconscious, it may be confusing even to analyze why a decision is difficult.

A second assumption concerns the content of unconscious ideas. A significant part of unconscious thought—and ensuing action—involves anxiety and efforts to avoid it in relations with others. In general, people experience anxiety as a sense of abandonment by others. Unconsciously, they may fear that others who could support them will no longer care for them, or they may fear punishment for some transgression. This study focuses on two types of anxiety. Guilt anxiety, or simply, guilt, involves a fear of moral punishment or exclusion for an offense against another person in violation of important values about social responsibility. For example, many bureaucratic workers wish to act aggressively against others and worry about the consequences. Shame anxiety, or shame, involves a fear of "loss of face" or total ostracism as a result of acting foolishly, ineptly, or somehow inappropriately. For example, many bureaucrats are concerned about not performing up to acceptable work standards.

A third assumption concerns the reason that such ideas are unconscious: simply, people attempt to defend themselves against the pain of anxiety by concealing unpleasant thoughts. They may do this in a number of ways. For example, someone who wants to act aggressively against another person but fears reprisal or censure may "repress" those hostile wishes. That is, he or she may retain the wishes and continue to calculate how to carry them out but push them from consciousness, relegating them to unconscious thinking. By concealing these wishes from awareness, the person can act without feeling hostile and does not need to worry about harmful consequences. Nevertheless, unconscious plotting may continue and even covertly dominate overt actions. Insofar as the scheming persists, the person may still feel anxious about it, although consciously may find this feeling puzzling, as he or she knows of no hostile wishes toward anyone.

Another significant defense against anxiety involves "denial," "splitting," and "projection." For example, someone may think ambivalently about another, with appreciation for past support but resentment about needing support. Whenever this person becomes aware of feeling resentful, he or she may become anxious about the consequences. The person may unconsciously defend him- or herself, first, by denying any hostile feelings toward the other person; second, by splitting these negative feelings off from the positive feelings; and third, by maintaining the appreciative feelings while projecting the angry feelings onto the other person. Consciously, he may reason in this way: "I do not resent that person; I really only appreciate him; in fact, if anyone really resents anyone, he's the one who resents me; and, come to think of it, it would be a good idea for me to act so as to protect myself from him." So long as this defense strategy takes place unconsciously, this person never needs to assess its basis in reality. The strategy does not eliminate the hostile wishes or the anxiety associated with them, but makes them seem like a realistic part of social relationships, with fears of reprisal more believable.

A fourth assumption concerns why adults with considerable formal education and life experience may, nevertheless, make such inaccurate assessments of social relations. A general explanation is that adults continue to experience anxieties they felt as infants and children and continue to defend themselves against them in ways that made sense in their earlier lives. Because they do so much of this unconsciously, they have difficulty recognizing when they think or act inappropriately. Adults may unconsciously "choose" to see other people in certain anachronistic ways because of a tendency to interpret new experiences through "transference" in the following way.

Because few persons or events are unambiguous, people must make some assumptions about them in order to act. The primary source for assumptions is past experience, conscious and unconscious. While people can draw useful inferences from past events, they may also unconsciously "transfer" assumptions from infantile or childhood experiences. For example, they may think of and react to contemporary organizational authority figures as if they were parents. Adults make such faulty analogies for two reasons. First, superficial resemblances between present and past experiences—for example, both supervisors and fathers exercise authority—seem to warrant thinking of the former in terms of the latter. In addition, strong childhood emotional ties to certain figures unconsciously encourage adults to conceive of current relations in terms that would re-evoke those feelings. Crucially, the reassuring emotional memories unconsciously defend the adult from the anxiety that might arise from thinking directly about contemporary experiences. For example, if someone can think of a boss as a loved parent, then he or she may avoid thinking hostilely about the boss and worrying about punishment.

These assumptions are elaborated in examining specific cases later, when other concepts are also introduced to interpret workers' actions.

Psychoanalytic Study of Organizations

The psychoanalytic study of organizational phenomena developed gradually from the original psychoanalytic concern with individuals. Freud's (1921; rep. 1959) main effort to examine organizational life concerned the church and the army. He indicated that group members unconsciously identify their ideals with a leader and then follow this person as the expression of their aspirations. Freud's view that collective phenomena are simply derivative from individual dynamics dominated the few psychoanalytic considerations of organizations that followed. Harold Lasswell, a political scientist who studied psychoanalysis in consultation with several of the early analysts in the 1920s, was the first to apply psychoanalysis to organizational and political matters per se. His studies (1960, 1963) of political leaders and administrators emphasized that people's attitudes toward power in their public roles express deep-seated feelings about power and authority in their early lives.

Wilfrid R. Bion's (1961) research with groups, beginning during World War II, was the first effort to use psychoanalytic methods and categories to study patterns of interaction in groups. Influenced by Melanie Klein's emphasis on the influence of social relations on personal development, he found that group experiences tend to evoke unconscious infantile memories and encourage collective re-enactments of certain typical early impulses. Much subsequent psychoanalytic study of organizations, particularly that associated with the Tavistock Clinic in England, has drawn on Bion to identify and interpret group activities within organizations (see Jaques, 1951, 1955; Menzies, 1975; Miller & Rice, 1969; Rice, 1958, 1963). This work builds on the view that organizations create situations to which people react in ways that are more than simply expressions of individual personality.

In the past 25 years the psychoanalytic organizational literature has grown considerably (see Hodgson, Levinson, & Zaleznik, 1965; Kets de Vries & Miller, 1984; Levinson, 1972; Levinson, Price, Munden, Mandl, & Solley, 1962; Zaleznik, 1965, 1967). But the cumulative literature is eclectic, reflecting the diversity of origins and the newness of this inquiry. Writers agree that individual unconscious processes affect actions in organizations and the actions of organizations. However, analysts disagree about the extent to which individual activity in organizations is simply an expression of individual personality and the extent to which it is a creation of the organization. Those who believe that organizations elicit original behavior disagree about whether the organization is a unique social setting or whether what is seen in organizations is generic.

The basic assumption about the importance of the unconscious has made some headway into nonpsychoanalytic organizational study. Merton (1940), although saying little about psychological dynamics, argued that conformity to bureaucratic norms could shape workers' personalities. Elaborating on this view, Hummel (1982) contends that fragmented organizational roles can lead to personality fragmentation. Thompson (1961) examined how working in hierarchical organizations creates anxiety. In different ways, both Crozier (1964) and Gouldner (1964) found that bureaucratic norms may be established to defend against and control anxiety. Argyris and Schön (1974, 1978), while avoiding direct references to the unconscious, demonstrate that what they call tacit habits of thinking make workers defensive in solving problems and find that organizational norms contribute to such defensiveness.

Comparisons with Other Perspectives

Still, the psychoanalytic focus on unconscious experiences, motivations, interests, and conflicts is exceptional in organizational study. In general, introducing consideration of the unconscious into analysis of organizational behavior makes actions appear more complex. In particular, both by pointing out certain self-defeating or antisocial tendencies in the unconscious and by identifying conflicts between these tendencies and conscious intentions, the psychoanalytic point of view emphasizes what might generally be seen as pathology.

For example, the book draws attention to behavior that is largely defensive, motivated by interests in avoiding shame or guilt toward organizational superiors. Thus a bureaucratic worker may approach an assignment with the primary concern of protecting him- or herself from others' finding errors, rather than directly attempting to find solutions for a substantive problem. The worker may warily collect and conceal information from others and may even refrain from challenging others' misinformation because of imagined, even if highly improbable, reprisal from others. Politeness may be more important than successful problem solving.

It may be said that such behavior is only one type of bureaucratic behavior. There are workers who generally act undefensively, rationally, and relatively successfully. Further, some may act defensively at some times and yet, under other conditions, succeed at rational action. The psychoanalytic perspective tends to neglect such relatively unconflicted, rational, successful action. Which type of action is more common, however, is a difficult question to answer. Observations of bureaucratic workers have been limited, and conceptual models of intrapsychic decision making are inchoate. Still, the psychoanalytic perspective would suggest that some relatively high proportion of bureaucratic behavior, as behavior elsewhere, involves the anxieties, defenses, and conflicts portrayed here.

Alternatively, there may be straightforward reasons for the wary ways just described. One might observe that defensive behavior tends to be a norm in organizations, regardless of individual personalities, and that, at the least, successful adaptation to an organization may require learning to act in typically defensive ways. Just as a visitor to a different country might attempt to learn the local culture in order to get along, so a new worker would adopt predominant patterns of organizational behavior. Many new workers, particularly those whose previous life experience has been restricted to the family and the classroom, find the entry into bureaucracy a "crazy-making" experience. Relationships are impersonal, and many seemingly unimportant events are recorded in writing with multiple copies sent to people near and distant. Some learning certainly is required for organizational integration.[2]

Nevertheless, one might ask why such learning seems for many to be relatively quick and enduring. Perhaps bureaucratic norms respond to common anxieties about authority and intimacy, including those which workers develop once in the organization. The psychoanalytic point of view would support a hypothesis of this kind, and Crozier's (1964) study of bureaucratic workers found evidence of such a relationship between bureaucratic norms and anxieties.

Still, one might argue that much defensive behavior in bureaucracy is both conscious and rational. Rewards are scarce, and other workers tend to act competitively. Few workers have the power to change either the structure of rewards or the behavior of others. Within these constraints, defensive action, even if regrettable, and even if sometimes a diversion from solving problems, becomes necessary in order to maintain one's interests. Indeed, defensiveness may be instrumental to problem solving. When there are differences of opinion and interest, it is reasonable to promote one's own position against others.[3]

This argument is valid. Defensiveness is appropriate in the face of real adversaries. The psychoanalytic perspective tends to be more concerned with conflicted actions and unconscious defenses and does underestimate the role of calculated, instrumental defensiveness. However, in any particular case one should ask what are the actual probable threats? One may make false, untested assumptions about others' beliefs and actions. One may exaggerate others' likely responses. A small chance of criticism may be blown up into a large chance of a reprimand, and the possibility of a reprimand may be distorted into fear of dismissal. Argyris and Schön's (1974, 1978) research uncovers numerous examples of bureaucratic defensiveness built on faulty assumptions. While defending oneself against harm may be appropriate, a realistic appraisal of the pending threat is necessary for mounting a realistic defense.

Still one other alternative interpretation may be offered to the perspective of analysis presented here, this one from within the psychoanalytic camp. It may be conceded that the analysis of unconscious motivations, conflicts, anxieties, and defenses is accurate, but there may be nothing peculiar to bureaucratic organizations about this. These anxieties and defenses are common in most or many people, and they are evoked in many other realms of life. The analysis of bureaucratic psychodynamics is only one area in the study of human psychodynamics. Is there anything peculiarly bureaucratic about what is observed in bureaucracy? In Zaleznik's (1983) terms, bureaucracy offers a *site* for the observed behavior, but does bureaucracy present special *situations* that contribute to the behavior?

Perhaps the most skeptical answer is offered by those who contend that bureaucratic organizations may be nothing more than psychic products (for example, Argyris, 1982; Diamond, 1984). People either directly create or tacitly endorse bureaucratic organizations and forms because these social structures provide defenses against personal anxieties. For example, bureaucratic structures minimize the likelihood of close personal contact among workers, protecting them against any anxiety that would be aroused by such intimacy. To be sure, bureaucratic structures create additional difficulties, but such anxieties, for example, are essentially secondary problems derivative from prior defenses against earlier anxiety.

Again, there is something to be said for this point of view. And yet one must be wary of reductionism, especially with a perspective that holds a deterministic, biological view of the origins of human behavior. There are probably some universal human psychological dimensions, whose pervasiveness makes them effectively biological characteristics, although it may turn out that the number of such dimensions is fewer than many Western psychoanalysts assume. These psychological characteristics may indeed influence all human action, including organizational ones. But this still does not explain all aspects of organizational behavior.

Contemporary bureaucratic organizations arose as a specific historical phenomenon, along with industrial capitalism. Bureaucracy was expected to regulate people's civic and economic activity in order that civil states might maintain order while economies grew. It has performed reasonably well in meeting these expectations, as well as in providing defenses against psychic anxieties. One might speculate that it is the relative success of bureaucratic organization in serving a combination of psychic and social functions that gives it such strength and persistence.[4] Reduction of bureaucracy to simply an economic or psychological function would deprive it of its richness and would deny many of the meanings it bears for its members.

The evidence in this book suggests that bureaucracy, as a distinct form of social organization, has its peculiar psychological consequences. Bureaucracy's contribution to human experience comes from the unique combination of three features: (1) it is an organization of work, with expectations about members' performances to benefit others; (2) workers frequently have insufficient authority to carry out their responsibilities; and (3) superiors who assign and evaluate work are often infrequently visible and accessible. The resultant psychological dynamics may affect the cognitive and political processes required for solving problems. At the same time, because these bureaucratic features are not simply psychologically or economically determined, organizational reform and the alleviation of existing psychological maladies may be possible.

The influences on actions in bureaucratic organizations are complex; this examination emphasizes the contributions of organizational structure. As noted, organizations, as well as their administrators, vary. In order to draw attention to apparent common effects of bureaucratic structures, the analysis gives some consideration to these real, important variations. In addition, individuals respond to similar organizational conditions in many ways. Developmental stage, sex, and personality orientation are three major sources of variation. While acknowledging the variety of responses, the book emphasizes a few as illustrative of bureaucracy's distinct effects on its members. Finally, the book focuses on problematic responses to bureaucratic structures, because they are widespread, and in order to call attention to the seriousness of the obstacles that bureaucratic structures pose to organizational work.

THE STUDY

Material for the book comes from an interview study of planners and public administrators. Practitioners in a relatively small sample (50 planners and 27 public administrators) were interviewed for two to four hours each in a semistructured interview. The subjects were drawn from the Maryland chapter memberships of the American Institute of Planners and the American Society for Public Administration.

Interviews focused on different aspects of bureaucratic work. Questions to the planners concentrated on the problem-solving process. They were asked what the ideal process would involve and what they contributed to it. They were asked how typical planning compares with such an ideal and why it may deviate. Finally, they were asked where they found satisfactions in their work and how they would change working conditions in order to improve planning and increase their satisfaction. Answering these questions required making explicit a cognitive map of the planning process and at least sketching the influences on stages in this process.

Interviews with the public administrators focused on the experience of power in bureaucratic work. They were asked about the relative certainty or ambiguity of their responsibility, authority, and accountability. They were asked whether they had sufficient authority to carry out their responsibilities. Questions were asked about sources of influence in the organization and their own ability to influence others and shape work. These questions required giving attention to personal relationships in everyday work. Although the substance of the two sets of interviews overlapped, in general interviews with the planners concentrated on intellectual requirements for problem solving, while interviews with the public administrators concentrated on organizational or political requisites.

The data from the interviews are qualitative, and they have been analyzed in ways that respect the speakers' meanings, partly by using direct quotations from workers as a way of identifying prominent cognitive and emotional topics. Both the sample size and the complexity of the phenomena limit the validity of any quantification. Where possible, an effort has been made to suggest the proportions of people who hold particular points of view or feelings. However, both the semistructured character of the interviews and the ambiguity of many issues sometimes make it difficult to refer to any group more specifically than by "most," "many," or "some." Although this imprecision is unsatisfying, the data do not permit greater specificity, and these general qualifiers suggest at least the magnitude of themes among responses.

A more detailed description of the study design, the sample, and methodological issues may be found in the Appendix.

SUMMARY AND PLAN OF THE BOOK

The chapters of the book are organized in five sections. The first eight chapters, as Table 1 shows, relate to four stages in the development of obstacles to bureaucratic problem solving, and the concluding chapter suggests ways to overcome these obstacles.

Chapter 1 provides the dynamic for organizational action by introducing the psychological contract. People approach organizations with expectations based on both reasoned, conscious analyses of their interests and skills and tacit, unconscious concerns about such matters as authority, aggression, and sexuality. Workers continually seek to negotiate satisfaction of these expectations in psychological contracts

Table 1. Summary of the Book's Analysis by Chapter: Stages in Development of Obstacles to Problem Solving

Psychological structure	Chapter 1: Workers' expectations of psychological contract
	Chapter 2: Pyramidal hierarchy of organization; ambiguity of tasks
Predicaments	Chapter 3: Ambiguous, unequal responsibility and authority
	Chapter 4: Invisibility of autonomous authority; shame and doubt
Responses to predicaments	Chapter 5: Escape and compensation
	Chapter 6: Ritualism; defensiveness; counterambiguity
Problem-solving approaches	Chapter 7: Scapegoating
	Chapter 8: Evasion; defensiveness; embrace of scapegoat role

with the organizations, and their success shapes their responses to others in the organization.

Chapter 2 presents the psychological structure of bureaucracy. A salient characteristic of bureaucracy is its hierarchical, pyramid shape. Control, or authority, is centralized, while responsibility is decentralized, and there is considerable distance between subordinates and those who have authority over them. At the same time, the bureaucratic structure is pervaded by ambiguity. This situation encourages workers to engage in transference in order to make sense out of their organizational relationships. Transference unconsciously introduces childhood assumptions and anxieties about authority into workers' perceptions. As a result, real bureaucratic difficulties are compounded by assumptions contributed from workers' psychic histories.

Chapters 3 and 4 describe the resulting predicaments that workers encounter as they attempt to solve problems. Chapter 3 analyzes the effects of working in situations in which responsibility and authority are ambiguous and unequally allocated. While ambiguity of responsibility encourages workers to do more, in order to obtain recognition, simultaneous ambiguity of authority and responsibility encourages workers to do less, in order to avoid blameful, punitive recognition. Workers may create "counterambiguity" as a means of extricating themselves from this predicament. In this way they attempt to undo the organization that does not satisfy their expectations, but they also lose the possibility of being recognized for success.

Chapter 4 analyzes the intricacies of working for an invisible, autonomous bureaucratic authority. This situation reinforces tendencies to transfer assumptions and feelings from other authority relationships and encourages attribution of great power to bureaucratic superiors. This transference confuses workers about the reality of bureaucratic authority. They respond to their perceived superiors with feelings of shame and doubt about their own performance, which prevent workers from acquiring power needed to solve problems and redress the inequality between their responsibility and authority. The doubt represents an effort to undo the organizational setting in which these experiences arise, but it also hinders acting to change the setting.

Chapters 5 and 6 describe workers' defensive responses to these predicaments. Chapter 5 examines workers' dramas—daydreams, plans for the future, or actual plays. These dramas may indict the organization for its failure to meet the expectations of the psychological contract, and the dramas aim to undo the organization and replace it with a more satisfactory one. They provide symbolic compensation, but they basically represent an escape from organizational problem solving.

Chapter 6 examines ways in which workers manage their formal assignments. They may seek to defend themselves by hiding behind technical or legal formulas, which they ritualistically invoke. In this way, they redress some of the inequalities of responsibility and authority that encourage dramas of escape and compensation. They also defend themselves from the risk of being blamed for problems. Despite the apparent firmness of technical claims, their effect is to introduce counterambiguity into problem-solving discussions: the issues never seem to be what colleagues believe they are. While counterambiguity is an effort to undo the political claims and intentions that threaten problem solvers, it also deprives workers of opportunities to exercise competence or be recognized for it.

Chapter 7 illustrates how hierarchical realities may be confused or obscured by transferred assumptions about the power of authority, with the result that attacks on problems are superseded by escapes from scapegoating. Magically solving problems through the symbolic sacrifice of potential advisors becomes more important than practically solving problems through collaboration. Scapegoating may be a worker's ultimate act of undoing the bureaucracy which prevents satisfaction of the psychological contract. Yet this symbolic effort and its possible success avoid testing workers' real ability and brings no possible solutions to real problems.

Chapter 8 describes responses to those who are asked to assist in solving problems under conditions when scapegoating is likely. First they may hesitate, but later may inadvertently or deliberately accept a scapegoat role. The practical and symbolic realms may become confused. Paradoxically, an advisor may embrace the scapegoat role as a defense against the difficulties of avoiding it. At the same time, an advisor may collude with a client in accepting the scapegoat role as a symbolic undoing of the organizational setting in which more straightforward advice giving and problem solving are not possible. The cost, nevertheless, is the advisor's loss of competence.

Chapter 9 offers guidelines for learning effective problem solving in bureaucratic organizations. Interventions are possible at each of the four stages portrayed in preceding chapters: the psychological structure, predicaments arising from the structure, responses to the predicament, and problem-solving approaches developing from these responses. Strategies may aim in each stage at four domains of change: individual thinking, feeling, assumptions, or actions; interpersonal relations; organizational structure; and societal structures or cultural norms. The book concludes with recommendations for intervention and discussion of future research needs.

NOTES

1. This is the process which Berger and Luckmann (1967) portray as "the social construction of reality."

2. For different perspectives on organizational acculturation, see Argyris and Schön (1978), Schein (1978), and Van Maanen (1977).

3. For two different views of "rational defensiveness," see Bacharach and Lawler (1980), and Downs (1967).

4. For an ambitious analysis of combined psychological and sociological meanings of bureaucratic organizations, see Mitroff (1983).

1

Problem Solving as Powerful Collaboration: A Framework

Social institutions influence the instinctual structure of the people living under them through temptations and frustrations, through shaping desires and antipathies. It is not a matter of the instinct being biologically determined, whereas the objects of the instincts were socially conditioned; it is rather that the instinctual structure itself . . . depends upon social factors. Unquestionably the individual structures created by institutions help conserve these institutions.

Every mental phenomenon is explicable as a resultant of the interaction of biological structure and environmental influence. Social institutions act as deterimining environmental influences upon a given generation. The biological structure itself has evolved from the interplay of earlier structures and earlier experiences. But how did the social institutions themselves originate? Was it not, in the final analysis, through the attempts of human beings to satisfy their needs? This is undeniable. The relations between the individuals, however, become external realities comparatively independent of the individuals; they shaped the structures of the individuals who then through their behavior again altered the institutions. This is a historically continuous process. (Otto Fenichel, *The Psychoanalytic Theory of Neurosis*)

THE SURPRISE OF BECOMING A BUREAUCRAT[1]

People take jobs for a number of reasons. Commonly, they make an assessment of their skills and interests, and they look for some fit between their potential and their images of various occupations.[2] The selection of an occupation is often not conscious, and it may be more or less restricted by social opportunities. There may be family traditions or expectations about what a son or daughter will do. In addition, labor market structures and economic recession may severely limit choices.

This book examines workers who have had relatively free choice about their occupations. They work as planners and administrators in government agencies. Their jobs are among the most skilled of the postindustrial, knowledge-oriented occupations. They are college graduates; most have at least one graduate degree. And yet relatively few set out to become "planners" or "administrators." These labels are nondescriptively general, but both convey an opprobrious hint of malingering or nonwork.

As these people moved toward the time of selecting an occupation, the names on college courses and the titles of graduate programs helped define their aspirations. Perhaps the planners were better served educationally than the administrators, insofar as "planning" has some commonly identified public characteristics and some claims to specialized technical skills. Thus planners are likely to have graduated from formal planning programs, while administrators have diverse educational backgrounds. For both, their occupations carried the comforting sense of continuing to be good, smart students. Somehow they would make a living at it. Somewhere there emerged the flattering notion that jobs in planning and public administration meant that someone wanted them as people who were valued and would give them influence in return for their intelligence. Though few ever precisely formulated it as such, the work they envisioned encouraged them to think of themselves as minor philosopher-kings.

These assumptions run through planners' thoughts about why they chose their work:

> Planning was attractive to me because it seemed to be dealing with substantive issues of the day that were important to me . . . the whole urban crisis.
>
> It was a way I could put together all my values for the cities. Social change . . . it was a place where you could have an impact.
>
> I wanted to deal with the problems of the cities. I saw it as a noble profession, as contrasted with being in business, selling Pepsi Cola, et cetera.
>
> I liked the fact that you could really make a monumental and historical change.
>
> Intellectually, it was challenging. I was trained as a generalist, and I could do a lot of things, and I could contribute to change. I was reading *Future Shock*. Planning was everything!

Planning and administration are bureaucratic work. And yet few of these people pictured themselves as bureaucrats. Few even gave much thought to the organizational setting in which they would work. Indeed, it is the gradual discovery of the surrounding bureaucracy that is most disturbing to planners; they feel that they were never properly warned.[3] Perhaps it is normal that people may think more about the intrinsic characteristics of an occupation than about the organizational setting in which members of that occupation work, but the bureaucratic character of planning and administration is an intimate part of those occupations. How could people be so surprised?

Most of these people say that they had not aspired to be bureaucrats.[4] When they had thought about future work and schooling, their aspirations ranged among the professions: law, engineering, medicine, teaching, architecture. A few—though not very many—thought in some way of going into business. These are all relatively high-status occupations. The highest status choices—medicine and law—provide the most autonomy. Their status may be due to the fact that the American ideal of the rugged individual striving against all odds is an important tradition, as is that of the rags-to-riches entrepreneur. Even if these myths have been intellectually dispelled, emotionally they continue to live. Thus it should not be surprising to find that people examined here did not give much thought to organizational surroundings. Indeed, what stands out in planners' confusion about their work is the conflict between their employment in bureaucracies and their earlier expectations of being entrepreneurial problem solvers—minds for hire, presumably at a good price.

Another reason why these people did not anticipate working in a bureaucracy is that the great public bureaucracies that employ them are recent inventions. Massive public agencies and programs were created as recently as the Depression. Thus little time has elapsed to assimilate such radical changes in the conditions of work and to transmit new expectations to workers. In particular, it was not clear how this transformation of work affected families' status aspirations.

The pursuit of status forms an undercurrent in the selection of an occupation. For example, when the planners were asked what attracted them to planning, many mentioned the estimated status of the occupation, possibilities of subsequent mobility into still better jobs, and frequently, the opportunity to exercise power over public affairs. An interest in status is not simply selfish. For workers such as these, status implies the opportunity to be recognized for what one is worth. They seek high-status jobs because they believe that they are highly competent, and they anticipate that their work will enable them to express this competence. Quite directly, planners said that they were attracted to planning by opportunities to use their intelligence in solving social problems:

> I had a lot of broad interests, and planning seemed to be the most possible of doing as many of all of these as possible. The fact that it is a very broad area of activity seems to be an area for constructive government, for dealing with policies and programs of government that affect services.

> It was a relatively new and growing profession in the 1960's, it was receiving a lot of emphasis in relation to the social programs of the federal government . . . and I think I saw it as an opportunity to put to practical use the social concerns which I got as an undergraduate.

> I felt guilty about the war. I had to do some powerful things of a creative nature to help other people. . . . I was not particularly attracted to the subject [of planning]. It was just to get going on it, and it seemed that the people there at that time were free-thinking, free-wheeling, an intellectual kind of flavor.

Work activities are inseparable from personal identity and are an intimate expression of an interest in being recognized as competent.[5] Such feelings undoubtedly distracted the planners and public administrators from noticing that they were going to become bureaucrats.

THE PSYCHOLOGICAL CONTRACT: A BIG DEAL

In order to understand the encounter between worker and bureaucracy, it is useful to think in terms of the negotiation of a tacit employment contract.[6] Most jobs are initiated with formal contracts, in which a worker offers a certain amount of time or achievement in return for a specified salary plus benefits, certain working conditions, and, perhaps, opportunities for advancement. Material interests are the substance of formal contracts. However, some of the strongest motivations for selecting a job concern intangibles, such as the opportunity to use one's skills in ways one deems competent, the possibility of having an impact on decisions, recognition for effort expended and results accomplished, and the authority to get others to do things in ways one regards as reasonable.

These motivations underlie expectations that workers bring to a job and employer but are rarely discussed formally, partly because these motives are difficult to define and thus elude concrete agreement. Sometimes they are not discussed because they are considered to reflect either idealism or selfishness inappropriate to a job. Many times the worker and potential employer may not trust one another sufficiently to address concerns that will affect the allocation of organizational power; each would prefer to attempt surreptitiously to pursue his or her own interests as time goes on. Only on rare occasions are such considerations openly discussed and negotiated. Then they may be incorporated in a formal contract, but, more likely, they become part of an informal understanding. More often, a worker may tacitly present such expectations, cloaked in other terms, and a potential employer may respond in a similarly indirect fashion. Both parties may assume that they reach an implicit agreement, but the covert nature of the discussion leaves doubts from the start whether there really has been an understanding and, if so, what it is about. Most often, there is not even a tacit contract struck.

Actually, it is unreasonable to expect to consummate any bargain about these interests at the beginning of employment. It takes time for workers, even unconsciously, to discover both how they might realize emotional interests on the job and then, whether and in what ways they might satisfy them. If an employer or supervisor thinks about such things, he or she also requires some time to identify their practical meanings in the organization. Further, interests in competence and identity, for example, unlike wages or hours, cannot be negotiated with single individuals; instead, they are worked out in relations with many other people who are not always clearly identified. The fact that such important agreements are not made at the same time as the formal contract makes it likely that a worker will give only intermittent, often passing, attention to the gradual working out of a deal on what may be some of the worker's strongest interests. In the end, no bargain on status or recognition, for example, is final. As working relationships change and as a worker's conscious or unconscious expectations change, deals are continually recast.

Some implicit contract of this kind is always negotiated between a worker and an employing organization. It is a "big deal," perhaps the strongest influence on a worker's response to an organization. A worker's assessment of the present and potential satisfaction of these expectations may lead, variously, to loyalty, hostility, or indifference.[7] To the degree that a worker perceives some satisfaction of these interests as attainable but in doubt, the organization has a hold on the worker, who will make special efforts to secure satisfaction. Because the social character of these interests makes them continually renegotiable, an organization may exercise a continuing hold over workers. Insofar as the satisfaction of these interests is ambiguous, still more is a worker likely to give the organization additional effort to obtain satisfaction.

Focusing on the peculiar ways in which this psychological contract is worked out in bureaucratic organizations, this book explores how planners and administrators see their emotional interests at stake, how they feel themselves to be threatened when they try to satisfy their interests, and what they believe they need to do in order to gain some satisfaction. As will be seen, bureaucratic workers' failure to find satisfaction leads them to take actions that conflict with their formal work responsibilities and organizational interests.

THE NATURE OF THE WORK: PROBLEM SOLVING AND POWERFUL COLLABORATION

Planners and public administrators tend to work in different jobs in different organizations, but much is similar in their work. It is helpful to begin with a description of the differences before focusing on what is shared. For example, it is customary to distinguish expertise as either substantive (knowledge about a specific field or problem area, such as health or employment) or procedural (knowledge about the process of analyzing and solving problems, regardless of the field). Public administrators are more likely than planners to claim substantive expertise, whereas planners are more likely to claim procedural expertise.

Substantively, most practitioners calling themselves planners—or city planners—work with problems involving the physical and economic development of cities or other relatively small geographic areas. For example, planners are likely to work in land use, urban design, transportation, sanitation, employment, and housing. Public administrators work in other fields, although this group may include planning administrators, who are also concerned with these matters. Public administrators are more likely than planners to be involved with human resources development in such fields as social services, health, education, manpower, and corrections.

Within their fields, planners and public administrators are likely to work in organizations with different procedural responsibilities. Most city planners work for local, regional, or state departments of planning.[8] These are "staff" agencies, whose purpose is to give elected executives advice on how to solve a variety of problems related to physical, economic, and, occasionally, social urban development. Unlike "line" agencies, planning departments rarely have responsibility for operating programs to implement solutions to problems. Their job is to develop plans and evaluate proposals in relation to those plans. Much of the time, planning departments respond to executive requests to evaluate proposals from the executive, other departments, or public groups. Although planning departments may lack the authority to implement programs, their recommendations or approval may be necessary before operating agencies can move ahead. Particularly when proposals or programs concern other departments, the boundaries between staff and line agencies may become blurred. This definitional ambiguity may lead to organizational conflict.

Public administrators work at all levels of government. Often, they are employed in line agencies, which carry out legislatively mandated programs in specific fields. A health department, for example, develops rules and regulations, budgets and allocates money, and organizes people's activities so that programs operate in response to legislative intentions. The department may also be responsible for providing services to the programs' beneficiaries. In addition, the department is expected to identify problems in the area of its responsibilities and to develop proposals that the executive may offer to the legislature.

As noted, planning agencies include both staff planners, and administrators who supervise and set agency policy. Similarly, within operating agencies, the roles of public administrators vary considerably, from top agency executives through mid-level managers to supervisors. The scope of public administration may also include subordinates who do not manage other people, but who manage information necessary for program implementation or work directly with clients who receive program services. The content of work varies within the organizational hierarchy. Toward the top, public administrators are more likely to be procedural experts rather than substantive experts. (For example, middle managers may be budget specialists or program planners.) As city planners, these program planners have advisory staff roles. The program planners differ from city planners primarily in the substance of the programs involved. In addition, the program planners in operating agencies have closer organizational links to implementation. These role variations thus introduce ambiguities into distinctions between planners and public administrators.

Hence it should be clear that, although there are differences between planners and public administrators, neither agency label, job title, nor even professional affiliation provides a certain indication of work role. These practitioners share an involvement in public problem solving, even though they differ in orientation toward either formulating or implementing solutions. A more important shared trait is that these practitioners work in bureaucratic organizations, whose peculiarly structured social relationships influence efforts to ask for and render advice about problems.

Most of the work of planners and public administrators is intangible and difficult to define. They talk with other people; they collect and analyze information; they write memoranda and reports; they arrange and attend meetings.[9] Although continuous, these activities produce few palpable results. Significantly, planners and administrators share public doubts about the value of their work, often floundering in efforts to define what they do.[10]

Why is this? Like other postindustrial occupations, planning and administration share two characteristics that deviate from traditional models of work.[11] First, the work and its products are intellectual, or mental, rather than physical, or manual. Unlike industrial workers who produce physical objects and changes in them, these service workers work with information and develop new relationships—new meanings—from it. Most commonly, these efforts are intended to solve problems. A solution to a problem first "occurs" in people's minds and only subsequently may be carried out in new social actions. And only planners and administrators who hold responsibility for implementation may become involved in the more tangible work of producing new social relationships and actions.

Planning and administration differ from industrial work in a second respect: they are social work. Industrial manufacturing entails work with physical materials; contact among workers is secondary, only required if it increases the efficiency of individual efforts. In contrast, service work is essentially work with other people; indeed, it cannot take place without it. Just as a physician cannot heal without a patient, a planner cannot begin problem solving without hearing from someone who has a problem and desires a solution. To be effective, working with others who have problems means giving advice.[12] Most commonly planners and administrators give advice to elected officials, their assistants, or public groups; at times administrators request advice from subordinates and consultants.

Intellectually, planning and administration involve problem solving. Socially, they are advice giving. The social character of their activities introduces a number of ambiguities. Often it is clear who identifies a problem and gives an assignment for work on its solution; at other times an assignment may be passed on by someone who does not know its origin or history. The problem assigned may be clearly described, but, more often than not, a situation may be only vaguely described. In this case, the would-be problem solver must begin by developing a coherent formulation of the problem. Thus not only the solution, but also the definition of the problem, must be satisfactory to the client giving the assignment. In this situation the person formulating the problem in an important sense "creates" the problem; it is no longer simply the client's problem.[13] This creative activity is necessary for effective problem solving; however, the advisor's formative role may lead a client to regard the advisor as literally the cause of the problem and to react accordingly. Further, the client may come to believe that the problem really belongs to the advisor and is the advisor's responsibility. Chapter 7 examines this possibility.

Each of these ambiguities is almost infinitely complicated by the fact that normally several parties are interested in any given problem. Even if only one party is a formal client, there are others who are concerned about whether a situation is defined as a problem, how it may be defined, and what action may be proposed to change the situation.[14] Differences may be the product of conflicting interests; what is a problem to some may be a source of benefit to others. Interested persons may have different information about a situation, or they may evaluate similar information differently. Rarely is it possible for a planner or administrator to eliminate such ambiguities simply by ignoring people. Their work requires negotiation. As much as possible, they must attempt to create agreement about how a situation is to be perceived. Further, they must attempt to negotiate a commitment to do something about whatever is troublesome about the situation. Only then is it possible to consider initiating some action that may solve a problem. Quite clearly, giving advice on solving problems entails collaboration—working together.

Centrally, collaborative action requires the exercise of power. In an important way, collaboration *is* the exercise of power—people create power when they work together to do things collectively that no one could do alone. Significantly, power brings gains and costs, and the latter may deter people from attempting to wield power, or even from thinking about it. Power brings control, but it also gives participants new responsibilities. Because power comes out of a relationship among people who struggle with and against one another to satisfy their interests, success requires that they care for one another when they attempt to act together. Caring involves not only an abstract intellectual commitment but also an emotional closeness in which parties reveal their needs and take others' needs seriously. This intimacy promises to lead to new agreements, but it also presents risks. For in order to create new possibilities, people must risk revealing needs that others may reject, as well as finding that they do not approve or support things which mean a lot to others. In negotiation, people must expose their commitments to groups, places, or ideals with the uncertainty of whether others will join and strengthen these commitments or decline and devalue the objects of the commitments.

Losing is always an unpleasant prospect, but even winning, and creating or expanding power, brings hazards. If someone becomes more closely tied to others, then they must all take sufficient care of one another in order to maintain their alliance. This may require time, money, or strategizing, but it also entails the worry that something may happen to jeopardize the coalition and lose the people or things that its power supports. In addition, those who prevail must take the responsibility for managing what they have acquired and the risk of failure. Even when success requires defeating others, victory is not simple. Then the winner must take responsibility for soothing or sustaining those who have lost.

In short, whenever bureaucratic workers attempt to give or take advice and solve problems, they must enter into close relations with others, with this dual potential for gain and risk.[15]

This concept of power as collaboration is not simply a wishful preference about how people should think about one another. It represents many people's basic unconscious beliefs about how they should relate to others in order to develop personally. Glass (1985) has discovered how deeply people make these assumptions about power in his research with schizophrenic patients. He has articulated the ideas of Frank, a recovering psychotic. Frank's past troubles reflect some serious idiosyncratic problems, but they also demonstrate the anxiety aroused by these common beliefs about power in human relations:

> For Frank, being powerful meant being able to sustain feelings of empathy and intimacy. Power necessitated encounters with others and the inevitable if sometimes painful recognition that it was impossible for the self to control all aspects of being. But the power to be empathic had to do with psychological development, the expression of emotional needs without being threatened by them or being afraid to respond to similar needs in others. . . . Frank rejected the argument that the defining elements of power are influence and manipulation. Although he accepted the existence of such operations in the social world, he simply was not concerned with power understood in these terms. Power for Frank was a process that binds the social order by lifting individuals out of their egoistic concerns. It was dependent on collaborative functions. If those functions were overwhelmed by self-seeking, then the entire basis of the social order was threatened. (1985, p. 220)

People espouse other concepts of power, most often the view that power is the superior organization of force in adversarial relations. This view is realistic in circumstances where social interests really conflict, and then it is sensible to think about strategies for defeating others. However, one reason why people often think of power in terms of hostile relations is to avoid thinking about the responsibilities and risks involved in the intimacy of collaboration. It is easier to put other people in the category of antagonists than to face the anxiety of trying to develop close working relations with them. Chapter 4 shows how collaboration requires greater psychological development than oppositional power relations.[16]

To be sure, bureaucratic advising does not involve the intimacy of relationships between family members or lovers. And yet most of those feelings, along with their attendant anxieties, lie near the surface of negotiations in problem-solving activities. A planner's responsibility for a community's economic welfare is not the same as the planner's responsibility for his or her family's welfare, and yet accepting responsibility for the community raises similar anxieties about whether one can offer adequate support, how one may be perceived if one cannot, and alternatively, what further obligations one will incur if one succeeds. Day-to-day contacts with organizational team members are not as intimate as daily contacts with a spouse or children, and yet teamwork carries similar anxieties about whether one can muster the authority required to meet one's responsibilities, whether one should risk conflict with others, how one may resolve any conflict, and whether, in the end, one will obtain or retain others' support and affection. In each case, action may bring both rewards and risks. In the same ways that responsibilities and intimacy in the family and friendships stimulate mixed feelings about action, so too do the possibilities of professional collaboration encourage ambivalence about acting powerfully with others.

Thus there are two sides to planning and administration. The rational, conscious side is concerned with collecting and organizing information in order to advise others on how to solve their problems. This work is complex, even ambiguous, but it may be studied and mastered. It forms the substance of textbooks on planning and administration, which emphasize their rationality and thus suggest that these tasks may be easy to carry out. Yet there is another side to planning and administration. The social relationships involved in advice giving arouse strong unconscious feelings. It may be inferred from their operation in family and friendship relationships that they may encourage irrational action, in conflict with all rational intentions. As we shall see, for many workers "saving face" or concealing their ignorance may become more important than solving present problems.

THE PSYCHOLOGICAL STRUCTURE: AMBIGUITY AND AMBIVALENCE

Most conventional theories of problem solving emphasize the rational side and give little more attention to the organizational context than planners or administrators did when they considered their work. Consequently, these theories neglect to take into account the ways in which both the social structure and, crucially, the psychological structure of bureaucracy make problem solving an ambiguous and ambivalent enterprise.

Most models of problem solving revolve around the optimistic assumption that as problems are caused by ignorance, they may be solved by greater knowledge.[17] This assumption suggests that problem solving is a rational activity.[18] There is no mention, for example, of conflict of interests. Instead, it is implied that disparate parties share an interest in understanding and finding remedies for conditions that bother any of them. What is required, then, is the organization of sufficient intelligence to collect and interpret information about these conditions. Collaboration becomes possible as people learn what they should do together.

There is some support for this rationalistic view. Particularly in situations where there is agreement about who is to be served and what values are to be applied to problems, rational calculation may contribute greatly to solving these problems. Additionally, many bothersome situations are so complex that reasoned analysis may go a long way toward elucidating what the problems might be and what might be done about them. Juvenile delinquency, spouse abuse, income maintenance, and institutional racism are examples.

Nevertheless, as these examples suggest, conflicts of interest also make it difficult to solve problems. People hold different stakes regarding which values should be applied to situations and whose interests should be served. Significantly, these societal conflicts are compounded by the effects of bureaucratic structures, which add to the difficulty of solving problems by creating new conflicts of interest among those with responsibility for thinking about others' problems. Members of bureaucratic organizations are divided in two directions, and each division sets workers against one another. Horizontally, workers are separated by a division of labor that assigns each a specialized task and responsibility. This breakdown of large projects into small pieces deprives most workers of any responsibility for collective, collaborative effort and encourages defensiveness over discrete domains. Vertically, bureaucratic structures tend to separate responsibility from authority, with responsibility more or less evenly distributed but authority concentrated toward the top of the organization. The creation of positions of responsibility without authority further encourages defensiveness by confronting workers with the risk of being held accountable for something they cannot control. Understandably, both divisions interfere with the communication, learning, and collaboration required for solving problems.

Still, such a political view of problem solving retains a notion of rationality. Whether one considers societal or bureaucratic politics, if one assumes that people are motivated by self-interests, then it should be possible to analyze those interests and anticipate the actions to which they will give rise. Problem solving becomes more complicated, but it is still possible to formulate the problems and find solutions to them.[19] Often it is also assumed that, even if parties conflict in particular situations, they all share an interest in continuing to find solutions for conflicts that arise. Today's adversaries, for example, may be tomorrow's allies.[20] This view extends the basic learning model by additionally requiring information about participants' interests and intentions. Insofar as these conditions hold, the main apparent obstacle to problem solving is any actor's reluctance to compromise, either because that actor regards its interests as irreconcilable with others or because it has so much power that compromise is unnecessary.

One major exception to all these conditions, however, seriously transforms the nature of problem solving. Periodically, people appear to act in self-defeating ways that conflict with their interests. Perversely, they may be creating problems for themselves, rather than solving them. When self-defeating actions are observed in private life, they may be labeled neurotic—they serve poorly understood unconscious purposes, and there is no obvious social benefit from them. Similar actions may often be found in planning, public administration, and other professional practice. However, they are rarely recognized, because their discovery would unsettle professionals' beliefs in their mastery. Significantly, the bureaucratic setting is a primary impetus to self-defeating action through the operation of its psychological structure.

This psychological structure, a distinctive pattern of feelings and emotional relationships, arises from the social relations among people who come together in an organization and attempt to enact satisfactory psychological contracts with one another.[21] The formal social organizational structure sets certain patterns for people's interactions. As people pursue financial reward, competence, power, recognition, and other goals, they react to this social structure in two ways. First, they may feel pleased with it or hostile toward it insofar as it satisfies or frustrates these aims. At least some of these feelings are often considered the basis for job satisfaction (see Herzberg, Mausner, & Snyderman, 1959; Robinson, Athanasiou, & Head, 1967; Vroom, 1964).

At the same time, relations with people who influence the distribution of money, power, or recognition may arouse more specific, though not necessarily conscious, feelings. For example, dependence on a supervisor's evaluation for a salary increase may unconsciously recall feelings about a father's control of a weekly allowance. Efforts to get recognition from a chief administrator may unconsciously remind a subordinate of earlier attempts to get parents' attention away from siblings. This process through which people may unconsciously assume that contemporaries are like others in their pasts, and through which people may feel about these contemporaries what they felt toward those others, is called transference and will be discussed in detail in the next chapter. Crucially, even though workers may not be aware of such feelings, they may influence workers' reactions to their organizations and their colleagues. Evidence for this may be found in the commonplace observation that bureaucratic authority is "paternalistic" or the complaint that bureaucratic rules make subordinates feel "childish."

Thus the organization provides a setting in which its members may seek economic, social, and psychic rewards and about which these people may have various feelings. Some emotional reactions, both conscious and unconscious, are widespread, whereas others are idiosyncratic. Always these responses are joint products of people's past histories and the organizational settings to which they bring them. The assumption underlying the concept of an organizational psychological structure emphasizes widespread emotional reactions; that is, any organizational structure may be expected to evoke typical emotional responses among members with similar life histories. Dominant emotional patterns in organizations may be said to be caused by the organizational social structure and may be considered to constitute a characteristic accompanying its psychological structure.

Atypical responses to an organization are also joint products of the setting and past histories. Exceptional responses may reflect unusual past experiences or may respond to marginal elements in an organization. These responses are analogous to deviations in the social structure.[22] Sometimes the deviance of reactions may be misleading, insofar as they may be intimately linked to the dominant feelings. For example, Chapter 7 describes how members of an organization may unconsciously create a deviant scapegoat role in order to serve perceived needs of the organization.

No emotional response to bureaucracy is necessarily unique to it, as its social structures may be found elsewhere. Bureaucracy is distinguished by the constellation of reactions associated with a peculiar arrangement of social relations.

Three features of the bureaucratic social structure dominate psychological responses. The first is the hierarchical organization of work, in which responsibility is dispersed while authority is centralized. Workers may lack authority necessary to carry out their responsibilities. Second, their work may be evaluated and rewarded by superiors who are only occasionally visible or accessible. Third, responsibilities, authority, and relationships among bureaucratic members may be ambiguous.[23] Thus workers may have difficulty being certain how well they are fulfilling their responsibilities and what their superiors expect when they make evaluations.

When workers lack authority needed to carry out responsibilities, they may feel frustrated about the possibility of doing competent work. The distance of supervisors injects uncertainties about recognition and resultant raises or promotions. Responses to these conditions may be overt and more or less conscious. Ambiguity introduces subtler, often unconscious reactions. Ambiguity in combination with the relative invisibility and inaccessibility of superiors may encourage anxiety about evaluation. The difficulty of knowing what supervisors want may lead subordinates to fill in pictures of their superiors from imagination.

Sometimes the psychological structure of a group or organization supports problem-solving efforts, or its effects may be so muted that it does not significantly influence these efforts. More often, the psychological structure interferes with attempts at substantive problem solving by throwing up interpersonal problems among the members. Concern about authority and the allocation of responsibility may command members' attention, with the consequence that substantive problem solving must be postponed until members can resolve conflicts about their relationships. Because these conflicts are normally unconscious, members are likely to deny their existence, and apparently rational approaches to substantive problems may founder atop unconscious struggles over authority.[24]

As a consequence, in situations where direct collaborative action may be appropriate to solve problems, these feelings may motivate bureaucratic members ambivalently, with conscious intentions opposed by unconscious aims: workers may want both to solve and not to solve problems (because other, interpersonal issues seem more important); to exercise power and not to exercise power (because it has possibly frightening consequences); to collaborate and not to collaborate (because there are no definite limits on what others want or will do); to take responsibility and not to take responsibility (because it may never be satisfactorily defined and fulfilled). With such complicated motivations, the straightforward problem solving implied by rational models may be an exceptional occurrence.

Such ambivalence—which is not unique to bureaucracy—helps to explain why people may act in ways that appear to be self-defeating. People have two sets of interests, which may be roughly divided into those which are conscious and those which are unconscious. The former are what are understood in normal discussions of political interests or self-interests, and they are commonly considered rational. The latter include a number of socially less acceptable interests, such as aggression, affection, domination, and submission. These rarely enter into awareness and frequently come into conflict with conscious interests.[25] Conventionally, these may be described as "psychological" or "irrational" concerns. Efforts to satisfy both sets of aims may lead to the satisfaction of neither, or efforts to satisfy unconscious interests may lead to the visible defeat of conscious, political interests. Thus as people respond to both groups of interests, actions may appear self-defeating, though it may be more accurate to say that some self-interests succeed at the expense of others.

Finally, under such conditions of ambiguity, anxiety, and ambivalence, the working out of a satisfactory psychological contract regarding conditions of bureaucratic membership and employment is likely to be extremely complicated and unstable.

NOTES

1. As noted in the Introduction, "bureaucrat" refers to anyone whose work involves bureaucratic elements.

2. This is the general finding of Holland's (1966, 1973) research. Within limits set by any constraints on choice, people tend to select occupations that are consistent with their personality orientations.

3. Some findings from my research on planners have been published in *Planners and Public Expectations* (Baum, 1983a). This and subsequent characterizations of planners in this introductory chapter come from and are elaborated in that book.

4. Levinson et al. (1978) made a similar finding among the private sector executives whom they studied.

5. Friedmann (1964) provides a good summary of personal interests and meanings invested in work. Sennett and Cobb (1972) present case material that illustrates both the

intensity of emotional investments in work and the psychic costs of not getting a return on these investments. Their case studies show what in this book is called the failure to enact a satisfying psychological contract on the emotional conditions of work.

6. Levinson et al. (1962) were the first to develop the concept of an informal psychological contract, which is presented here.

7. Presthus (1978) portrays three typical patterns of response to perceived organizational incentives: upward-mobile striving, indifference, and ambivalence. Hughes (1958), followed by Kramer (1974), focuses on the "reality shock" of discovering that personal work expectations cannot be satisfied by an organization. Cherniss (1980) analyzes the conditions under which this disappointment may lead to "burnout" and withdrawal from the organization.

8. Approximately one-third of planners work as private consultants (Getzels & Longhini, 1982). Their clients may be private individuals or corporations. However, even public agencies may engage consultants when permanent staff lack time or specific skills.

9. The general and often social character of planning and administrative work presents a sharp contrast to professional emphases on specialized technical skills adhering to this work. For summaries of planners' activities, see Hemmens, Bergman, and Moroney (1978) and Schön et al. (1976). For a description of managers' activities, see Kaufman (1981) and Mintzberg (1973).

10. For planners' examples, see Baum (1983a).

11. For representative discussions of changes in work associated with postindustrialism, see Bell (1973), Heilbroner (1973), and Hirschhorn (1975).

12. On differences between goods production and service production, see Baum (1981).

13. Forester (1981) and Seeley (1967) have clearly explained the ways in which formulating problems entails taking and creating problems.

14. Mitroff (1983) has presented an ambitious survey of the "stakeholders" in organizational problems.

15. Myeroff (1971) provides a good phenomenology of the relationships involved in caring.

16. Among political theorists the concept of power as collaboration is associated with Arendt (1958). As noted, most other definitions refer to the ability of one actor to obtain certain behavior on the part of another, often explicitly against the will or desires of the other. Even if the results prove to be mutually beneficial, the power relationship is considered asymmetrical. This concept of power is evident among such widely divergent authors as Bacharach and Lawler (1980), Dahl (1961), French and Raven (1959), Kipnis (1974), Salancik and Pfeffer (1974, 1977), and Wrong (1980). As Chapter 4 will suggest, this view of power may be especially prominent in contemporary Western society. As suggested already, people may prefer it because it provides a defense against the uncertainties and anxieties of collaboration.

Significantly, those who study problem solving endorse Arendt's concept of power as practically necessary for creativity. It is essential, they argue, for people to be able to work with others to perform actions that no one can do alone (see Delbecq, Van de Ven, and Gustafson, 1975). Argyris and Schön (1974, 1978; Argyris, 1982) shed light on the ways in which people tacitly make adversarial assumptions as a defense against the anxieties of the collaboration required to solve problems.

17. For a critical summary of this literature, see Etzioni (1968).

18. Different writers have placed different emphases on their interpretation of the limited capabilities of human intelligence. Moynihan, for example, received notoriety for insisting that a limited understanding of the problems of poverty called for a period of "benign neglect" by the federal government. His conservative assessment of the War on Poverty is presented in Moynihan (1970). Simon, on the other hand, has formulated a concept of "bounded rationality" that retains an emphasis on rational action while acknowledging limits to information and understanding. His most recent presentation of this view is Simon (1983).

19. These premises underlie game theory. For a critical review of that literature, see Mack (1971). Wildavsky has developed these assumptions most articulately in a concept of political rationality. See, for example, Wildavsky (1979). Recently these assumptions have been applied in the growing field of dispute resolution. See, for example, Fisher and Ury (1983), Susskind, Bacow, and Wheeler (1983), and Susskind and Weinstein (1980–81).

20. However, there are good reasons for observing today that such an assumption does not hold, that there is a "constitutional crisis" in social problem solving and conflict resolution. See Baum (1977, 1983a).

21. In writing of groups, Bion (1961) characterizes the phenomenon as the "group mentality." Emphasizing the remarkable, binding consensus that members tacitly create, he describes group mentality as:

> the unanimous expression of the will of the group, contributed to by the individual in ways of which he is unaware, influencing him disagreeably whenever he thinks or behaves in a manner at variance with the [group's] basic assumptions. It is thus a machinery of intercommunication that is designed to ensure that group life is in accordance with the basic assumptions. (Bion, 1961, p. 65)

22. Weick (1979) emphasizes that an organization should be considered a process wherein individuals form and sever temporary, partial relations with one another. In such a "loosely coupled system," similarities in people's aims and interests lead to overlapping organizational goals and regularities in organizational structures, but many idiosyncratic, loosely regulated niches are inevitable.

23. For a provocative discussion of ambiguity in bureaucratic organization, see March and Olsen (1976), Perrow (1977), Thompson (1967), and Weick (1979). Chapter 3 examines ambiguity in detail.

24. Representative examples of the influence of this psychological structure on group or organizational problem solving include Bion (1961), Hirschhorn and Krantz (1982), Hodgson, Levinson, and Zaleznik (1965), and Rice (1963).

25. It is not simply that unconscious interests do not enter into awareness which may lead to conflicts with conscious interests. The problem is that those interests which become unconscious were repressed because they were socially unacceptable. Their intolerability is the source of conflict with conscious interests. See Freud (1920; rep. 1977).

2

Confusion About Power in Bureaucracy: The Influence of the Psychological Structure

Because I have a problem with internal politics, I could be totally wrong about who has power, because I tend to shut out who has power in the organization. But it is one thing which I need to overcome if I want to advance in the organization. (A federal agency worker)

Effects of the psychological structure are seen in confusion about the nature of organizational power. The ideal bureaucratic structure should strictly regulate members' relationships and preclude the personal exercise of power. However, workers often report that their relationships are ambiguous and involve an exercise of power that they cannot definitely attribute to anyone. Although ambiguity may be a source of opportunity, this perplexity incapacitates. An important consequence is that as workers cannot understand the character and requirements of power, they will be unable to collaborate and solve problems.

A paradoxical relationship exists between the bureaucratic social structure and its psychological structure. Although the social structure is intended to restrict individual discretion and personal power, members' personal needs intrude in two ways to create an emotionally charged psychological structure. One way, already mentioned, is that workers may want to control other people for various purposes and use the organizational structure as a means toward these ends. Thus formally regulated relationships may take on emotional meanings as vehicles toward highly charged personal goals. The second way in which people may transform the impersonality of the social structure is more covert. Unconsciously, people have difficulty imagining a world devoid of human actors.[1] As a consequence, whenever they see actions, they impute personal motivations to them. The less explicitly personal an action is, the more an observer needs to draw on his or her imagination to determine who is responsible for the action, what the person's reasons must be, and crucially, how the observer feels about that person.

Both these processes thus defeat intentions to design organizational social structures to operate impersonally. Insofar as organizational social design does not take members' personal motivations into account, there is little chance that these motivations will be consistently satisfied or regulated by the social structure. Opposition of the psychological structure to the aims of the social structure is more or less certain.[2] This chapter describes that process.

THE WEBERIAN IDEAL-TYPICAL BUREAUCRACY

The principles underlying common bureaucratic social structures are articulated in Max Weber's "ideal-typical" bureaucracy. In constructing the ideal type, Weber emphasized the logical requirements of a prototypical bureaucracy in order to explore their implications. His model provides a standard for analyzing and comparing actual organizations.

The ideal-typical bureaucracy is characterized by a regularity of actions and relations that ensures predictability of both the ends and means of organizational activity. Work is impersonal and nonpolitical; it is governed by rules rather than personal needs, and people take no actions that are not intended to lead directly to the production of formally designated organizational ends. No one has interests or intentions that conflict with organizational goals.

The first principle of bureaucracy, Weber wrote, "is the principle of fixed and official jurisdictional areas, which are generally ordered by rules" (Weber, 1967, p. 196). Formal rules define specific roles for workers and specific domains of work; these role expectations constitute the worker's responsibility. In addition, these responsibilities are accompanied by appropriate authority, or ability to get others to act in ways needed to carry out the responsibilities:

> The authority to give the commands required for the discharge of these duties is distributed in a stable way and is strictly delimited by rules. (p. 196)

Moreover, workers in each role have explicitly delineated relationships with workers in other positions:

> The principles of office hierarchy and of levels of graded authority mean a firmly ordered system of super- and subordination in which there is a supervision of the lower offices by the higher ones. (p. 197)

Authority is systematically varied from higher to lower levels of the bureaucracy, and lines of accountability between those in higher and lower levels are fully set forth.

Generally, the management of the bureaucracy is based upon written documents. The files of the organization present permanently stored, unchanging precedents that clearly identify the fixed rules governing members' actions. Weber emphasized that

> The management of the office follows general rules, which are more or less stable, more or less exhaustive, and which can be learned. (p. 198)

Thus in the Weberian ideal-typical bureaucracy, for any situation a worker's role, responsibility, authority, and accountability are completely set forth. Because general rules so thoroughly govern the actions of organizational members, any occupant of a position may be expected to act in the same manner as any other.

This predictability specifically precludes the exercise of power. Members work in response to delineated individual expectations and incentives. Moreover, authority in the ideal-typical bureaucracy is based on expertise, with conditions for advancement expressly specified in terms of work performance, rather than position, personality, or political organization.[3]

The consistency of this model results from Weber's endowing it with all the characteristics logically required by the essential aspects of bureaucracy. However, another reason for the model's consistency is that Weber concentrated on a sociological description of bureaucracy. He did not consider workers' personalities or the possibility that their interactions might constitute a psychological structure at least as influential as the formal social structure.

REAL BUREAUCRATIC LIFE

When bureaucratic workers describe their own organizations, they mention a number of deviations from this logical model. Workers interviewed in this study report that, although formal rules govern a wide range of actions and relations, these rules may be operational only if endorsed by otherwise influential actors. In addition, actions may be directed or constrained by unwritten rules that influential actors consider important. Even when positions appear to be bounded by rules, effective limits are set by people who interpret the meaning and applicability of rules. Not only does the efficacy of rules depend on how workers interpret them, but, more generally, actions and relations are not so impersonal as the ideal-typical model suggests. Responsibility is defined by interpersonal relationships. Influence depends not simply on technical competence, but also on personality. Individual effectiveness rests on the ability to establish good relations with other workers.

Moreover, workers say, bureaucratic organizations are political systems in which people seek to acquire power. Workers are interested not simply in organizational productivity, but also in the enlargement of their personal domains. In the service of their own personal interests, they organize others by manipulating definitions of responsibility and authority, developing informal systems of rewards and sanctions, and influencing decisions about organizational goals. In these ways workers not only strive for ends that diverge from formal organizational goals, but do so by adopting strategies inconsistent with those that are formally prescribed.

Taken together, these observations sketch a bureaucratic organization that differs significantly from the Weberian ideal. Personal motivations and interests in power operate alongside, even in opposition to, formal regulations. Organization theorists since Weber reflect these workers' feelings when they refer, directly or indirectly, to ways in which typical patterns of thinking or feeling cause organizations to deviate from the impersonal ideal type.

Phenomenological and open systems theorists have called attention in different ways to the social constructedness of organizations, their impermanence, and the permeability of their boundaries. They observe that people have many needs, that the investment in work organizations, while intense, is only one of several, and that people may attempt to use organizations to satisfy aims developed elsewhere, such as in earlier life or in the contemporary family (see Bennis, 1966; Burrell & Morgan, 1979; Hummel, 1982; Katz & Kahn, 1978; Mitroff, 1983; Weick, 1979).

Human relations theorists have emphasized the importance of people's needs for competence and recognition in leading to the creation of informal groups.

Through these groups workers set their own performance standards and evaluate and reward one another for meeting them, even when group expectations conflict with organizational norms (see Argyris, 1960; Herzberg, Mausner, & Snyderman, 1959; Katz & Kahn, 1978; Likert, 1961; Mayo, 1933; McGregor, 1960; Roethlisberger & Dickson, 1939).

Students of organizational politics observe that workers acquire interests in controlling others in the organization, not simply to accomplish organizational aims, but also to gain a sense of personal control. Although most of these writers normally say little about unconscious motivations for power and take organizational politics at face value, they regard strategizing for turf control as axiomatic (see Bacharach & Lawler, 1980; Downs, 1967; Mintzberg, 1983; Pfeffer, 1981; Pfeffer & Salancik, 1974; Salancik & Pfeffer, 1974).

Students of planning and administrative processes have observed that information collection is often incomplete or haphazard and that communication is frequently limited or distorted. In general, these writers focus on the characteristics of information, rather than those of the people who collect and transmit it. For example, they emphasize cognitive difficulties in managing large amounts of information and the cumulative problems arising in communications through any extensive network of people (see Cyert & March, 1963; Etzioni, 1968; March & Olsen, 1976; Simon, 1976). However, more psychodynamic explanations come from others who observe that fear, defensiveness, and tacit, if not unconscious, intentions and habits hamper problem-solving efforts (for example, Argyris, 1982; Argyris & Schön, 1974, 1978).

How a possibly unconsciously shaped psychological structure develops and operates in conflict with the intentions of the formal social structure is illustrated in the following case material.

A WORKER'S EFFORTS TO UNDERSTAND
BUREAUCRATIC POWER

How did the psychological structure arise, and how does it affect people? It is useful to begin this inquiry by examining how bureaucratic workers refer to the structure. Significantly, many bureaucratic workers find it difficult to be certain whether they are limited more by personal power or by impersonal rules. For example, a federal worker, asked to speculate about the balance of power and rules in his work, offers a contradictory analysis:

> I have to live within the rules. What I do is consistent with the rules. The ability to interpret the rules [matters]. But the fact of the matter is, I think I am limited ultimately by . . . I am not sure. . . . People are enforcing the rules. But if they are not enforcing the rules—in this sense, it is both [rules and people], but, in fact, the rule is probable. But I don't have any rules saying that I can't do this. It depends on what the initiative is. If it is consistent with the rules, then the people will be there. If you are talking about the rules of the organization, sort of the unwritten rules, that brings on a different kind of question. . . . I think that even if nobody says this is a rule, you need to think about if this is acceptable.

His uncertainty about the balance of impersonal rules and personal influence is measured in the number of times he qualifies an observation with "but." He is confused about whether formal rules or personal relations matter more, but he suggests that their connection itself is ambiguous.

In part, such confusion may result from superiors' deliberately ambiguous portrayal of their actions as a strategy to control subordinates. Another federal worker reveals this in describing his own power:

> [Power in this organization] depends on the individual. Some individuals have personalities that are such that they exercise the power personally. . . . The word processing man puts most things on a personal basis. When he exercises his power, it is on a personal basis. We try to make things part of the process, so that personality does not get involved. If a new organization is not approved [for a license by our section], it is not approved because of this regulation or that rule. Then it is not personal. He realizes that you cannot approve it because a rule stands in the way. [What about persuasion?] One of the things which persuasion involves is, you build a certain relationship with a client, so that he is convinced that you have done your best. He may not agree with all your recommendations, but he will go halfway to try to meet you.

Thus the administrator's actions contribute to confusion over whether rules or personal power matter more, but it is important that the administrator's strategy may work precisely because it responds to an existing ambiguity about the effective organizational governing structure.

The following worker's discussion of organizational power indicates that for some people this ambiguity may reflect unconscious feelings about the bureaucratic social structure. These feelings then make up part of the psychological structure and influence subsequent actions. This worker is a federal employee, a staff assistant to a deputy secretary. She has been in her current position only six months, but she has worked in the agency for five years and held other bureaucratic jobs previously. She is in the early stages of her career, but she calls herself "a committed bureaucrat."

She was asked to describe the basis of organizational members' power: who has it, what each person has power over, and what is the source of each person's power. When she talks about power, she cannot be certain whether it is personal or impersonal. Her confusion sheds light on the development of a bureaucratic psychological structure:

> Because I have a problem with internal politics, I could be totally wrong about who has power, because I tend to shut out who has power in the organization. But it is one thing which I need to overcome if I want to advance in the organization. . . . Power: let me say this, although I don't have any traditional sense of power. I don't have any authority, but you talk in terms of cooperating. I am pretty sure that when I call anybody in [a departmental program office], I am pretty sure that they will call back. When I call anybody in another agency, I express myself as the Secretary's staff. If they only knew how plebeian I am. I use the Secretary's power. I think name-dropping is the appropriate word. But it is real. But I don't abuse it. . . . Most of what I see is . . . impersonal, appropriately impersonal. Structural power. If you start to play the political things, which may be the more personal aspect of power. . . . But I shut them out. I am not aware of any negative use of personal power having touched my life. But it could have been, and I am not sure I would have recognized it. I have seen some personal manipulation . . . some people who have avoided the civil service system and gotten somebody hired. . . . Power should be based on formal lines of authority. Power should not be based on whether I like you because we play racquetball together. . . . Because I believe that the system can be a positive one, I am willing to give the benefit of the doubt. Innocent until proven guilty.

   This description is filled with conflicting statements: the bureaucratic organization is personal, but it is impersonal; actions are political, but they are not political. The description is syncretistic, combining anecdotal descriptions of personal influence and political action with the Weberian language of impersonality and nonpolitical action. What explains such an apparently contradictory view of bureaucratic organization?

The Legitimacy of the Weberian Ideal Type

Bureaucratic workers daily see personal and political actions. Because these actions influence their work life and their careers, workers naturally would want to mention them when discussing bureaucratic organizations. However, workers have difficulty finding a language that readily describes bureaucratic organization in terms of these experiences. The only legitimate language available is that of the Weberian ideal type, even though it does not recognize some of the most important daily experiences.

   The belief that the Weberian ideal type represents the legitimate model of organization is evident in the comments quoted. The worker acknowledges exercising power but explains that her actions and those of others are "impersonal, appropriately impersonal." She adds that power should be based on formal lines of authority. Thus, although her personal experiences are inconsistent with the Weberian model, the Weberian language of impersonality is invoked as the legitimate description of bureaucratic organization.[4]

   In this interpretation, bureaucratic workers who offer ambiguous or contradictory images of bureaucratic organizations may be considered to be confused by the dominance of the legitimate Weberian language and the absence of an alternative language that would give recognition to the personal and political phenomena that workers see daily. If workers had access to a legitimate alternative language, they could be expected to develop views of bureaucratic organizations that accurately, consistently, and inclusively reflect the personal and political activity of bureaucracies.

The Ambivalence of Personal Experience

However, a careful reading of the last quotation suggests that the ambiguous image of bureaucratic organization reflects more than a difficulty in finding a language that accurately and legitimately represents experience. The worker appears to want to find the conditions of the Weberian ideal type, even when her experiences seem to show something else. The ambiguity of her description appears to reflect her ambivalence in the experience of personal influence and power in the bureaucratic organization, as suggested by the number and content of "but" statements.

   At the beginning of the extract, she comments that she tends to block from consciousness any consideration of who has power in the organization but that she needs to overcome this hesitation if she wants to advance. Then, having implicitly given herself permission to look at who exercises power, she goes on to describe her own power. She observes that she doesn't have any "traditional" power or any authority but that she does have power to force cooperation by using the secretary's name. She follows this observation with the statement that this "name-dropping" may seem unimportant, but it is real power. Yet just as soon as she has acknowledged that she may have power herself, she begins to qualify this admission. She notes that her power is real, but she quickly adds that she doesn't abuse power. Next, once having opened up the exercise of power for discussion, she begins to reflect on the possibility that power may have been exercised over her. The recognition of this possibility leads to a denial of the presence of power. She notes that she and others may use personal power in a political way but says that she wouldn't know, because she ignores power and politics. She concludes by observing that she has not noticed that others have used power against her for harm, but she adds that she is oblivious to power anyway.

This worker concludes with the same obliviousness to personal power with which she began, but in the process of momentarily recognizing power she makes two important statements. She says that, in fact, she exercises power over others. In addition, she allows that others may exercise power against her, perhaps to her disadvantage. This opening and closing of the mind's eye to the personal exercise of power expresses considerable ambivalence about personal power. She recognizes personal power and then denies that it exists.

In order to understand this complicated description of power in a bureaucratic organization, it is useful to analyze what the worker denies. Having talked about the personal aspect of power, she denies abusing power and then moves on to refer to possible negative uses of personal power against her by others. Apparently she makes an association between her own exercise of power, which may be used to abuse others, and others' exercise of power against her: the personal exercise of power may bring punishment. Thus the denial of her own exercise of power seems to be really a denial of the possibility of others' exercising power against her to punish her. Power is seen as aggression, which brings retribution in kind.

In this context, the assertion that most power is "appropriately impersonal" appears to be more than merely an intellectual endorsement of the Weberian ideal type of bureaucracy. The assertion really seems to be an expression of a wish: personal power is aggression that brings punishment; therefore, any power exercised must be impersonal, and so harmless and able to go unpunished. This wish has two components. If one exercises power, then one should be considered the nonaggressive instrument of impersonal power, with the result that one's efforts to control others are not punished. And if another worker attempts to exercise power over one, because one has not exercised aggressive personal power, the other's exercise of power is not aggressive and not harmful. In a bureaucratic organization impersonal power controls without inflicting harm.

But where does the threat of harm—or punishment—come from? Possibly it comes from the worker's experience of being a subordinate. For all her talk about exercising personal power, she emphasizes that she works under the secretary, she uses the secretary's name rather than her own for anything important, and she regards herself as "plebeian." She experiences herself primarily as a subordinate; being controlled and dependent is the essence of her position. This condition may give rise to a wish that the control be benevolent. For some reason she does not trust other people. She believes that if the control over her were the result of personal actions by others, then these others would seek to abuse and harm her. Therefore, she wants to believe that the control over her is impersonal. For this reason, others' control or power is "appropriately impersonal." She elaborates that the power that controls her is "structural," although she does not explain how either the motivators or instruments of "structural" power could be other than fellow workers. Thus control should be impersonal in an organization in which other people cannot be trusted.

But this does not explain her mistrust of other people. The worker may find bureaucratic dependency unsettling: it may unconsciously recall memories of infancy, in which she lacked control over any of the external world and in which her very survival depended on another's goodness. This recollection, unconsciously transplanted into thinking about adult work relations, may arouse anxiety, conveying a warning of annihilation.[5] People normally respond to anxiety with anger toward the perceived source and with aggressive wishes to remove it.[6] Bureaucratic superiors who evoke such memories may thus become the objects of hostility. Because a worker recognizes that control is ultimately exercised by other persons, this wish to remove the control is an aggressive wish to remove—and perhaps harm—another person or persons.

In fantasy this wish could give rise to concern about retribution, experienced as anxiety in connection with any anger toward superiors. Workers may defend themselves against such anxiety by doing two things unconsciously. First, they may deny, or disavow, the anger. Then they may project the same anger onto those who exercise control in the organization. That is, they may unconsciously assume that the anger they feel is really someone else's anger. Still, once someone feels angry toward a superior about being controlled, in fantasy the aggressive wishes may be carried to destructive success, and the worker may feel guilty about imagined harm to the superior. Now the subordinate's anger, which has been disowned and attributed to the superordinate, may be experienced as the vengeance of the superior-as-conscience. The worker's perception that others feel angry toward him or her does not remove anxiety but does give justification to aggressive impulses toward others, these impulses now reinterpreted as self-defense. Unconsciously, this act of projection would produce feelings of mistrust toward others in the organization.[7]

Support for this analysis comes from a comment the federal employee made about her accountability to her supervisor:

> I am accountable to my supervisor for everything. Because so much that I am writing goes out under somebody else's name, *I have fairly much had to suppress my own identity* in this job, because I rarely get to write in my own name. [emphasis added]

Part of this statement is a complaint that others get credit for work she does; this concern is examined further in the next chapter. Significantly, she says that her very identity is in jeopardy as a result of relations with her supervisor. Fear of losing one's identity is the essence of anxiety, and this remark suggests that she refers to unconscious feelings of anxiety in her relation to her supervisor. She may feel anxious because of childlike "accountability for everything." Significantly, however, she does not say that her supervisor takes her identity from her; rather, she suppresses it herself. She annihilates herself, actively choosing the fate she fears most, in order to avoid retribution for acting out resentment about dependency against her superior-as-conscience.[8]

Consistently, her insistence that power is impersonal may be seen as an effort to deny the threats conjured up by feelings of resentment about being in a subordinate organizational position. If the controls holding the worker subordinate can be considered impersonal, then there is no evident risk of punishment. This interpretation would explain the woman's final comment that in a bureaucracy there should be no punishment, because the worker, as anyone, must be considered "innocent until proven guilty." In short, this worker describes power ambiguously because she is ambivalent about exercising it. Although she may need to assert herself in order to execute her responsibilities, she feels that assertion carries the risk of retribution, and her anxiety about punishment discourages clear thinking about power.

STRUCTURAL CAUSES OF AMBIVALENCE

Different early life experiences make some persons more ambivalent about power than others. This woman's special concern about being controlled undoubtedly reflects idiosyncratic infantile and childhood events. Nevertheless, similarities between her mixed feelings and those of other bureaucratic workers point to ways in which the organizational social structure embeds ambivalent tendencies in the psychological structure.[9]

The influence of bureaucratic social structures on feelings about power rests on the way in which power is exercised in bureaucratic organizations. Because bureaucracy is designed to permit relatively few to supervise the work of many, subordinates experience power in a peculiar way. On one hand, a supervisor exercises evident authority by giving assignments and evaluating work. At the same time, because the supervisor has responsibility for many workers and in turn reports to another supervisor, the supervisor may be only intermittently visible and accessible to a subordinate.

Some technologies require closer supervision than others, some supervisors may have discretion to spend more time with subordinates, and some may choose to concentrate their attention on subordinates in order to have influence with their own superordinates (Blau & Scott, 1962). In general, however, the more subordinates for whom a supervisor has responsibility, the less likely he or she is to have contact with any individual subordinate. Particularly in large organizations, even if an immediate supervisor is relatively accessible, important decisions about the purposes, methods, and evaluation of work may be made by someone higher up who may be only infrequently visible.

Moreover, when supervisor and subordinate do meet, they do not communicate much about themselves; instead, they focus on work. The supervisor's primary responsibility, for which he or she is evaluated, is implementing organizational goals, and discussions with subordinates emphasize instrumental tasks. Even if a supervisor is interested in subordinates' private concerns or feelings about the job, this primary responsibility makes it difficult to give these matters much attention. If private problems interfere with organizational work, a supervisor may refer a subordinate to an employee assistance program. However, even here the supervisor's main concern is getting work done, restoring the worker's productivity, and someone else, a specialist, gives help.

In addition, not only is a supervisor rarely available on demand, but when the supervisor observes subordinates, he or she may do so without asking permission or recognition in response. Thus a subordinate may find that the supervisor is both relatively anonymous and self-sufficient, or autonomous. The supervisor may seem able to satisfy his or her own needs without any assistance from the subordinate. A supervisor's relative anonymity arouses a subordinate's curiosity and anxiety about the superior. The addition of relative inaccessibility and autonomy, making subordinates unable to control their supervisor's presence, intensifies the anxiety. Together, these conditions may unconsciously remind adults of infantile anxiety about the invisibility and inaccessibility of their mothers.[10] These feelings add to a supervisor's formal authority and make a subordinate ambivalent about exercising power.

Reification, Transference, and Ambivalence

Someone who is powerful but anonymous and indifferent poses a riddle: this person is significant, as someone with authority, and yet reveals little about him- or herself. Who is this person who exercises such authority? What does he or she want? What may be safely assumed about this person? One way to contain the anxiety expressed in these questions is to develop definite images of the person in authority and to attribute specific motives to him or her. However, in an ambiguous situation the only certainties possible are reifications, or concrete conceptualizations. Although conversations, observation, and even gossip provide clues about the personal characteristics of the authority and the relationship with it, they still leave puzzling gaps. Thus imagination is the main source available for filling in missing information.

This process of sketching in the details of authority, the process of transference mentioned earlier, has two elements. First, the most compelling reservoir of ideas for imputing traits to authority is infantile and childhood experiences with authority. Those experiences tend to form templates for conceptualizing later events. Because early relationships with parents and other family members were so crucial to development, any dissatisfaction with these relationships impels one to work them out in later, seemingly similar, relationships. Second, most of this cognitive and emotional inference from infancy and childhood is unconscious, partly because some events took place before the conscious mind developed and partly because many events are too upsetting to remember straightforwardly. As a result, people fill in details of adult authority in ways that elude their conscious awareness.

Such transference is a universal, everyday process. Bureaucracy especially encourages it because important relationships there are particularly anonymous. The formal influence of a superordinate is reinforced in a subordinate's mind by the unconscious influence of childhood memories, which, because they operate unconsciously, contribute invisibly to the superordinate's authority.[11]

The specific features unconsciously added to bureaucratic authority vary with members' past experiences. However, the dominant characteristic of infantile and childhood experiences with authority is of being small, powerless, and dependent on someone large and powerful.[12] For example, some of the remarks of the woman quoted earlier hint at this when referring to her relationship with the departmental secretary:

> When I call anybody in another agency, I express myself as the Secretary's staff. If they only knew how plebeian I am. I use the Secretary's power.

When she refers elsewhere to her responsibilities, she speaks in a way that is bureaucratically conventional but which also echoes childhood relations with parents: "Everything that I do is always going to somebody higher than myself."[13]

Maccoby (1976) reports that few of the corporate managers whom he interviewed said much about their childhood or even remembered it clearly. He speculates that childhood experiences of powerlessness remain in unconscious memories and may dominate reactions to adult work situations. These executives rarely see conscious connections between work and childhood, he suggests, because

> they don't like to think of childhood because it was a time in which they were helpless, powerless. Perhaps their lack of freedom and self-determination now keeps them from remembering and experiencing feelings of childhood because they are too painful and too similar to work-related feelings of powerlessness that are constantly being repressed in the present. (1976, p. 159)

Bureaucratic workers may rationalize transferred feelings into plausible, though reified, images of their supervisors. Common memories of parents may make supervisors appear especially assertive, dominant, and powerful. Subordinates' reactions to these perceptions vary with early life experience. Some may be angry about being dependent on such authority. They may feel unconsciously recalled childhood anger toward parents. They may also feel adult resentment about feeling so much (remembered and transferred) childhood dependency on adult authority such as supervisors. Such dependence may not simply seem inconsistent with respectful adult relations, but it may also appear to block initiatives required for assignments. The supervisor may appear aggressive and imperious to these subordinates, threatening to take away their power and competence.

Others may find the supervisor's power reassuring. They may feel anxious about risks in making independent choices in conflictful situations. Perhaps they fear displeasing authority by making unacceptable choices. Giving in to the imagined wishes of a dominant, omniscient authority will rescue them from this anxiety. In the end, whether workers react positively or negatively, their feelings about authority, even if tied to inaccurate perceptions of the supervisor-subordinate relationship, help to delineate an otherwise poorly defined relationship.

These views of a supervisor's power may contribute to ambivalent responses to the exercise of power. On the one hand, both the accomplishment of formally prescribed work and the development of a respected domain require a worker to influence others in order to solve problems. Yet the very idea of influencing—exercising power in relation to—other workers may evoke anxiety about offending the supervisor, who in fantasy is seen as controlling all subordinate action and as punishing anyone who is insubordinate. Thus a worker may have instrumental reasons for wanting to exercise power vis-à-vis other workers, including some superordinates, and yet may experience anxiety about taking an action not explicitly authorized by a superior. This ambivalence may become rationalized in uncertainty about responsibility and authority, as the next chapter illustrates.

Ambivalent feelings about exercising power may contribute to ambiguity about the nature of power in bureaucratic organizations. On the one hand, every supervisor represents a reality in which a bureaucratic worker is, in fact, subject to the power of others. At the same time, an apparently powerful supervisor may seem largely invisible. This near invisibility may create uncertainty whether power is exercised or, if it is exercised, whether it is somehow impersonal, or "structural." Common sense may lead a worker to think of the power of a supervisor, who has a face and a name, as personal power. However, the imagined dangers from the personal power of a supervisor—retribution in response to aggressive wishes or actions—may discourage the worker from thinking of the supervisor's power as personal. Consequently, an accurate assessment of the nature of power is influenced less by a weighing of empirical evidence than by a desire to avoid anxiety that may be associated with the specific finding that power is personal.

Thus the bureaucratic social structure encourages the creation of a psychological structure through transference. Infantile and childhood ambivalence about parental authority become superimposed on the adult's analysis of organizational authority and may make appropriate professional initiatives seem threatening, with resulting confusion about power.

PSYCHOLOGICAL AND PRACTICAL CONSEQUENCES OF AMBIGUITY AND AMBIVALENCE ABOUT POWER

This confusion about power has major practical consequences by causing workers to invest at least as much effort in not acting as in acting. First, confusion impedes the will to act. Anxiety about punishment for the exercise of power leads workers to spend considerable energy on maintaining reasons not to act in ways that could bring retribution. Workers may vehemently insist that no reasonable initiatives are possible. To bolster this claim, they may assert that the organization is indecipherable. As a result, in order to develop a clear will to act, a worker must expend extra energy to overcome these defensive arguments against action, as well as to redefine the situations that evoke the anxiety requiring such protection. In short, when a worker has recurrent concerns that the action necessary for solving problems may also be a form of aggression that brings punishment, the worker must invest time and energy in untangling this connection before unambivalent action is possible.

Second, so long as thoughts of action make a worker anxious, perceptions of the organization are unlikely to be clear. As noted, a worker may insist that the organization is unintelligible as a defense against acting. Even if the worker does not do this, the effort invested in avoiding risky initiatives detracts from the time and energy required to clarify an ambiguous perception sufficiently to identify potential courses of action. The case example here suggests that a resulting implicit organization theory is likely to be syncretistic—combining a recognition of personal motivations and political interests with a Weberian denial of the personal and political.

Third, insofar as the worker cannot decode the organization, the worker's attention and effort are especially susceptible to influence by others, notably supervisors. Without a clear picture of the organization, subordinates will have little ability to assess the reasonableness or feasibility of others' requests. Even though subordinates may have necessary formal authority to act, they will have trouble understanding the scope and type of actions they can or should perform. Thus the bureaucratic psychological structure encourages conservatism.[14]

Finally, for all these reasons, bureaucratic workers will have difficulty collaborating powerfully with others in order to solve problems. They will have trouble identifying their personal and political interests, as well as the interests of those affected by the problems they attempt to solve, and they will have difficulty acting on the interests they uncover. Not only is their personal influence diminished, but the influence of their clients and constituents suffers as well.

The next section looks more closely at two predicaments that the psychological structure presents to bureaucratic workers. Each interferes in different ways with acting powerfully to solve problems.

NOTES

1. Evidence for this statement comes from observations of infants and psychoanalysis of adults. The infant's earliest model of the world is its body. Other human beings take on meaning to the infant only as they satisfy various bodily needs. In turn, early examples of human relationships—primarily with parents, secondarily with siblings, then with others—become terms of reference for all subsequent experience. For examples and analysis, see Fenichel (1945), Ferenczi (1916; rep. 1980), Mahler, Pine, and Bergman (1975), and Winnicott (1971).

2. This point has been made from the beginning by the human relations school of organizational behavior (see Argyris, 1960; Likert, 1961; Mayo, 1933; Roethlisberger & Dickson, 1939). This book differs from most human relations writers in emphasizing workers' unconscious motivations as well as their conscious personal needs.

3. Crozier (1964) has noted that both the classical theory of bureaucracy and its human relations revisions have rejected power as an issue. He writes

> If one believes [as the classical view holds] that co-ordination, conformity to orders, and the will to produce can be brought about with only economic and financial incentives—i.e., if the world of human relations is ignored altogether—then power problems need not be taken seriously. But the exact reverse is operating within the human relations approach. If one believes that a perfect equation between satisfaction and productivity can be achieved under permissive leadership, one does not have to study power; one has only to fight to accelerate its withering away. (p. 149)

4. The language of impersonality may not only be inappropriate for describing some bureaucratic experiences, but may also interfere with understanding deviations from an impersonal model, as the language discourages personal responsibility and reflection on experience. See Hummel (1982, Chapter 4).

5. Anxiety represents a signal of potential danger experienced by the ego from the threatened breakthrough of unconscious impulses into consciousness—and the ensuant pursuit of these impulses in overt action (Fenichel, 1945; Freud, 1936; rep. 1963). In infancy anxiety is related to potential conflicts between instincts and external objects, such as other people. At this time the imagined consequence of the enactment of unacceptable sexual or aggressive impulses is annihilation. As images of other persons are internalized, anxiety may arise from conflicts between instincts and these internal representations, for example, formulated as the superego (Freud, 1946). As a result, in later life the imagined consequence of enactment of unconscious impulses is recast in socially more appropriate terms. Thus Sullivan (1953) characterizes anxiety as "anticipated unfavorable appraisal of one's current activity by someone whose opinion is significant" (p. 113). In the cases of the workers discussed in this book, anxiety represents a signal of potential disapproval, abandonment, or aggression with which a worker imagines that others will respond to his or her imagined actions.

6. Some people say that they find dependency on external control to be comforting because it provides security. However, it is important to ask why these people find security in external control. Winnicott (1965) suggests that people may get two types of security from dependency. When people doubt their own goodness and their ability to control destructive impulses, subordination to the control of another defined as good unconsciously provides protection from the possibility of actually harming others. When people fear retribution for aggressive impulses toward others, such subordination may unconsciously provide protection. Thus it should be clear that the security matters only because of the dangers. Indeed,

organizational research has produced some consistent findings. For example, Crozier (1964) concludes that people create impersonal hierarchical relationships as a defense against face-to-face relationships and risks of personal dependency. Argyris (1982) argues similarly.

Further, in order to enjoy the security offered by subordination to external control, it is necessary to accept a position of weakness and childlike dependence. Most adults find this condition uncomfortable (Bion, 1961). Thus it is possible that many who say that bureaucratic control provides them security share the same conflicts about control as those who resent it, but the two groups defend themselves against these conflicts differently. For an analysis of different responses to bureaucratic subordinacy, in addition to that emphasized in the text, see Zaleznik (1965).

7. For a discussion of the process of projection of aggressive impulses, associated anxiety, and guilt, see Klein (1948) and Riviere (1939; rep. 1964). Such a projection of anger onto a supervisor would help to account for two common experiences of bureaucratic workers. One is workers' feelings that their supervisor does not like their work or may not trust them to do competent work. The other is workers' amorphous fears of punishment by a supervisor for any type of insubordination.

8. Zaleznik (1965) characterizes this response to subordinacy as "masochistic"—someone seeks to evoke aggression from an authority figure as a means of guarding against his or her own aggressive tendencies. Zaleznik argues that this pattern is likely when a subordinate ultimately prefers being controlled by another to controlling others and actively seeks to become so. In distinguishing this response from others, Zaleznik points to the influence of personality orientation on workers' actions. The analysis in this book demonstrates that this response is most likely when subordinates attempt to take initiative in the common bureaucratic situation of lacking authority to implement responsibilities.

9. Although this book emphasizes the influence of bureaucratic social structure on ambivalence about power, it is important to recognize other, often complementary, influences. For example, organizational culture, which includes interpretations of the social structure, conditions feelings about power. For an analysis of the interplay of structure and culture on these feelings, see Menzies (1975). In addition, the technology of work, insofar as it shapes social relationships, affects responses to authority and interpretations of power. Some indications of the influence of technology may be found in a comparison of Gouldner's (1964) study of coal miners and Hodgson, Levinson, and Zaleznik's (1965) study of psychiatrists. This book plays down the possibility that ambivalence about power is simply neurotic, primarily because the response is apparently widespread. Nevertheless, individual differences in experiences with authority certainly condition responses to the exercise of power. See, for example, Davies (1980) and Presthus (1978). The style of organizational managers or leaders influences responses to power by shaping and interpreting organizational structure. In turn, managerial style may be understood as an expression of managers' personalities, organizational structure and roles, and work technology. For a discussion of influences on and psychological responses to managerial styles, see Cohen and Cohen (1951), Fiedler (1965), Henry (1949), Kaufman (1981), Lindgren (1982), Maccoby (1976), and Mitroff and Kilmann (1975).

10. See Bowlby (1973) for an analysis of children's responses to the invisibility and inaccessibility of their mothers. See Memmi (1984) for an examination of adult responses to the undependability of those with authority.

11. For a discussion of the process of transference in clinical relationships, see Brenner (1976), Fenichel (1945), Greenson (1967), Orr (1954), and Racker (1968). For studies of transference in organizational relationships, see Hodgson, Levinson, and Zaleznik (1965), and Kets de Vries and Miller (1984).

12. Few children experience a single authority figure. Father, mother, and an older sibling may all exercise authority. Each of these might be a source of transference material for adulthood experiences. Hodgson, Levinson, and Zaleznik (1965) illustrate the ways in which these childhood authority figures may give rise, respectively, to reified adult images of a

paternal-assertive superordinate, a maternal-nurturant superordinate, and a fraternal-permissive superordinate. The selection of transference material in adulthood will depend on the subordinate's idiosyncratic childhood experiences, such obvious personal characteristics of a superordinate as sex and age, and any evident distinctive patterns of behavior on the part of the superordinate. The discussion of transference in this book emphasizes relations with a superordinate who is perceived to be paternal and assertive.

13. From the relatively brief interview alone it is impossible to be certain that these comments refer unconsciously to parental relationships. At the least, this language fits equally well both bureaucratic and parent-child relationships and provides evidence that workers may view the experiences similarly. Citations in later chapters provide additional evidence for workers' transference.

14. This dynamic explanation for commonly observed bureaucratic conservatism contrasts with a more linear view offered by Hummel (1982) in an analysis of bureaucratic language. Hummel argues that the everyday language used by bureaucratic workers tends to be built on a logic of analogy, rather than one of causality. Workers' internalization of the logic of this commonly used language, he suggests, contributes to an uncritical attitude toward authority and purpose in bureaucratic organizations. Such an attitude reinforces what he identifies as the intentionally conservative purpose of bureaucracy as an organization of human effort. Certainly Hummel is correct that bureaucratic language affects workers' actions. However, this case study offers two modifications to his argument.

First, the study suggests that the conservative influence of bureaucracy is psychodynamic as well as cognitive. In addition, the study suggests that the relationship between bureaucratic workers' language and their actions may be more complex than Hummel indicates. The language of the worker quoted includes not simply the analogical syntax that Hummel emphasizes; the language also includes an explicitly causal discussion of personal and political relations. The analysis of the case suggests that the syncretistic juxtaposition of the two types of language—the personal and political apparently challenging the impersonal and apolitical, but never evidently with legitimacy or security—contributes to an ambiguity that interferes with criticism. If bureaucratic language were exclusively analogical, acausal, impersonal, and apolitical, then each apparently personal or political action would offer the basis for clear criticism. Edelman (1977) makes this argument in analyzing the consequences of holding contradictory beliefs about the causes of social problems. The ambiguity of the beliefs together, he suggests, permits acceptance of otherwise unacceptable conditions. Marcuse (1969) makes a similar argument in analyzing the repressive social consequences of the political tolerance of both orthodox and apparently critical ideas.

3

Ambiguous and Unequal Responsibility and Authority

I don't want to be recognized. People have been shot down so many times that they want to sit back at their desk and be left alone. You have a lot of people here who know their job, but they are not allowed to do their jobs. (A career civil servant)

Authority and responsibility are unequally distributed in the bureaucratic hierarchy. Responsibility—what a worker is expected to do—may be relatively evenly assigned throughout the organization. In contrast, authority—the ability to make effective requests of others, including the assignment of responsibilities—is concentrated at the top. As a result, bureaucratic workers tend to lack the authority necessary for the completion of their assignments. This predicament is a common complaint.[1]

When inequalities of responsibility and authority are experienced in the context of relatively anonymous authority, workers find choices of action laden with ambiguity, intensifying the predicament of acting with their responsibility unequal to their authority.

AMBIGUITY IN THE BUREAUCRATIC STRUCTURE

Workers commonly portray bureaucratic situations as ambiguous. A national study of bureaucratic workers found that ambiguity is pervasive in organizations because workers normally receive only incomplete information about their task requirements and expected responsibilities (Kahn et al., 1964). A survey of policy makers regarding use of information in decision making revealed that many do not feel involved in decisions because they find both their responsibilities and their role to be ambiguous (Weiss, 1980).

Public agency staff members regularly note this. One reason may be that the division of labor is fluid. As a federal worker remarks:

> There is a certain level of ambiguity in terms of our division. The Office of Special Projects . . . the Secretary also has an administrative assistant. So there is some overlapping. I may get asked to do a speech, but then her A. A. gets asked to do a speech. For the whole division there is overlap.

The division of labor may be fixed, but guidelines may still be unclear, as a career state bureaucrat observes: "The broad outlines are quite distinct, but the actual tasks

involved change from day to day." Even when responsibilities are explicitly designated, they are subject to interpretation. A staff attorney notes,

> We literally sign a contract. We have critical and noncritical standards that we have to
> meet as supervisors. [And yet] nothing is very clear, because in this business everything
> is subject to interpretation. No two attorneys as investigators will agree on every-
> thing. . . . It is not a role situation.

A caseworker in a state agency comments that rules may not be definitive:
"What happens sometimes is, we don't get clear or precise instructions from [the
top]." Further, even if a subordinate takes initiatives, administrators frequently react
unforeseeably. A state planner remarks,

> You do a plan, and that means that you get involved with everybody, but there is a lot
> you don't find out, because there is an inner circle.

A local budget officer explains,

> Ambiguities lie in the area of, sometimes, the interpretation of the law. Sometimes
> anticipated interpretations of the law by the judiciary or someone. A limitation on
> authority is whether or not the right people are going to go along with your decision,
> whether or not there will be people who would try to overturn what you do. And the
> probability of their success.

Sometimes it is not even clear who makes final decisions. A state policy analyst, after
first sighing, then laughing, says,

> I have the authority to pull together anything that I want and show it to my supervisor.
> Because everything that I do is always going to somebody higher than myself, it always
> goes to my supervisor. In this sense there is no authoritative authority.

Such ambiguity in responsibility and authority represents a major violation of
the conditions of the ideal-typical bureaucracy. Decision theorists observe that these
deviations make the choice of actions itself more ambiguous. Whereas members of
the ideal-typical bureaucracy may plan in the light of definite goals and await highly
probable outcomes, workers in conditions such as those just described often have
difficulty selecting actions because even their purposes are ambiguous.

Some ambiguities in responsibilities may be the result of the complexity, uncertainty, and changeability of bureaucratic tasks. It may be difficult to define fixed
responsibilities in a way that will be appropriate to all tasks. Some responsibility
may be deliberately kept ambiguous in order to encourage collaboration. Under
these common conditions, past experience may provide little guidance, because its
meaning may be obscure. Notions of organizational and individual purpose may be
so vague that they do not help with specific problems. Even self-interest may be so
ambiguous as to fail to provide direction for a simply selfish notion of responsibility
(Cohen & March, 1974; March, 1978; March & Olsen, 1976). At the same time,
some members of an organization may attempt to manipulate others' perceptions of
their responsibility in order to encourage them to do undesirable work or to refrain
from valued work. Gouldner (1964) observes that supervisors frequently create
ambiguity about responsibility in the belief that resulting anxiety and insecurity
effectively motivate performance.

Some reasons for the ambiguity in authority are similar. In general, ambiguities
in relations among organizational members may blur the character of power, as the

preceding chapter indicated. The complexity of tasks may confuse people about who has relevant expertise. Even if authority is meant to be allocated according to a formal hierarchy, workers' differences in competence, interests, and ambitions lead to ambiguity in specific situations. Moreover, members may acquire informal authority that gives them influence over not only how decisions are made, but also which issues are open for deliberation. Nevertheless, because decisions may not be made in neat, clearly identifiable ways, influence, while real, may remain obscure.

Observers note that ambiguity may be a source of opportunity, in which loosely defined roles, responsibilities, authority, and accountability permit imaginative, active organizational members to respond in appropriate ad hoc ways to poorly defined but pressing problems (see Bennis & Slater, 1968; Michael, 1973; Schon, 1971; Weick, 1979). March (Cohen & March, 1974; March & Olsen, 1976) has articulated such advice into a "technology of foolishness": workers should be playful with regard to organizational rules and humble concerning expectations of themselves. They should experiment with roles, they should entertain unlikely interpretations of situations, and they should accept the probability that actions will come to mean something other than what they expected.

In referring to the flexibility needed with ambiguous tasks, a department head quickly notes:

> It is a matter of how you use the opportunities you have. But there is some ambiguity. . . . You could build a rule book that would cover you in every case, but you would lose your flexibility. There has to be ambiguity in the situation. The guidelines are not always clear.

A fiscal officer reports, "Sometimes I like to have it nebulous, so that I can come back with something that is do-able." For a state division head, ambiguity is the essence of opportunity:

> I go ahead and do it. Authority is not the issue, in the sense that only the treasurer of the organization can sign payroll checks. Ninety-nine per cent of it is judgment, exercising common sense. The problem is that we get too hung up over whether we are in charge. Get on with it.

A clerical worker in a county office presents an intriguing account of her own uses of ambiguity to get needed jobs done:

> Authority is given to me by the organization. I don't have authority but what is given to me by my supervisor. I do have the authority to give someone an account number, to tell them that they have money to spend. I can communicate with the State, and my boss will back me up. We argue a lot. . . . I guess that is about all the real authority that I have. Sometimes you take more authority than you should in order to get something done. I do that a lot. For example, I had a check to pay. I wrote the check and took it to the fiscal officer and told him to sign it, and it took him all day to sign it. I circumvented a couple of people. It is the old story of knowing who as well as what.

Still, these examples are exceptional. Although ambiguity may present opportunities for creative action, for most people it inhibits acting. Along these lines, March and Olsen (1976) observe that, when confronted with ambiguity, most people allocate their attention according to relatively simple notions of role, duty (responsibility), and obligation (authority). Several studies suggest that ambiguity discourages action by causing workers to feel anxious, to see others as more powerful, and to doubt their own efficacy. People respond to perceptions of both real and imagined, even if exaggerated, differences in authority (Cohen, 1959; Kahn et al., 1964; Weiss, 1980). Combined ambiguities and inequalities in responsibility and authority are important elements in the bureaucratic social structure. This chapter examines the resulting predicaments for workers and identifies typical psychological responses.

AMBIGUOUS RESPONSIBILITY

Ambiguous responsibility and ambiguous authority each contribute to distinct problems, and bureaucratic workers react to the problems differently. Workers respond to ambiguity about responsibility with efforts to make responsibility more certain. However, they respond to combined ambiguous responsibility and authority with efforts to make responsibility more ambiguous. Thus different elements of the same situation contribute to opposite impulses. The dynamic explanation for this ambivalence rests with workers' experiences of threats to their identity and their anxiety about recognition.

Threats to Identity

Bureaucratic workers believe that they do good work. They often worry that ambiguities about responsibility will obscure credit for their work and, worse, erase the public identity associated with their efforts. One common concrete expression of this concern is the complaint that written work frequently does not bear the name of the person who performed it and, indeed, often carries someone else's name. An analyst in a state agency observes, "Our names don't go on the report. The division's name goes on it."

Frequently work goes out with the signature of a supervisor. Although it is possible to take some private satisfaction from knowing that a supervisor uses one's work, often it is uncertain whether, when, or in what ways a supervisor does use the work. This uncertainty is described by a special assistant to a departmental secretary:

> As a staff person to the secretary, you are responsible for all the work you do. The end result is that the work you produce comes out under the name of the secretary. I really don't know how much of my work goes out without the secretary being involved in it.

For some workers this experience may simply be disappointing or frustrating. Others regard it as a direct attack on their identity. One federal worker states that "the biggest adjustment" to her present agency

> is that I can't sign my correspondence. I needed to have five people sign something I wrote. It took away some sense of authorship and the time it takes to move a piece of paper. . . . I found it very frustrating. I was accustomed to doing something myself.

Another federal worker, quoted in the last chapter, complains,

> I am accountable to my supervisor for everything. Because so much that I am writing goes out under somebody else's name, I have fairly much had to suppress my own identity in this job, because I rarely get to write in my own name.

The suppression of personal identity is described with unintended irony by a high-ranking member of a federal department in response to the simple question, "What do you do?": "Right now I am not doing anything. I am doing assignments for my boss." No matter where one puts the emphasis in this statement, the result is the obliteration of a personal identity as a competent worker. If the "not doing anything" is emphasized, it means this worker is simply idle, wasting time. If the "I" is stressed, it means that, indeed, work is being carried out, but this worker has no part in it because it belongs to his boss. The consistency with which workers refer to the loss or suppression of their name, authorship, or identity is clear evidence that such comments express deeply felt concerns about what difference they make in the work of the organization and what they gain from their efforts.

Working for Recognition

One way to make peace with these concerns may be humble surrender, assuming that recognition will inevitably be settled upon others and expecting that one will have little public identity as a competent worker. This self-effacing position is espoused by a career bureaucrat:

> The job needs to be taken seriously. The person does not need to take himself seriously all the time. I didn't understand this originally. Don't get caught up in self-importance.

Even so, a state planner acknowledges that this position may be unsatisfying:

> I probably don't look for [recognition] so frequently because I don't give a damn at times, because if you look for it and don't find it, you just get grumpy. It is just the nature of the job. Your basic question of why.

Few bureaucratic workers can accept the view that taking themselves seriously is a mistake. If responsibility is ambiguous, it must be clarified to give credit to those who do the work. This conviction is clearly expressed in bureaucrats' most common response to a question about what they look for as indicators of their effectiveness: "recognition." Both planners and administrators emphasize recognition and compliments from colleagues as more important criteria for effectiveness even than impact outside the organization.[2] To begin with, they would like their efforts to be acknowledged and praised.

Some doubt they will ever get recognition. Skepticism about getting meaningful rewards from the organization is expressed by a state worker:

> I do so many things. I'm not sure if I would ever put my job at the top of the list of things that make me happy in life. Somebody saying, "This is a good job, a good piece of writing." Some sort of recognition, yeah. Personal measure, I don't know. I know when I am doing it that I am doing the best I can.

In order to secure recognition, he attempts a partial emotional divorce from the organization. He distinguishes work for the organization from work for himself and weighs the latter more heavily. Further, while agency supervisors may evaluate and recognize—or not recognize—the work for the organization, he reserves for himself the prerogative of assessing the work that belongs to him.

Nevertheless, there is some ambiguity in his solution, as he indicates in the last sentence that he regards some work for the organization as also work for himself and attempts to hold up his own evaluation against that of supervisors. This suggests that the resolute self-effacement of the worker may contain some rationalization.

Significantly, the continuing ambiguity about boundaries between organizational and personal work sustains the organization's hold on the worker. It is difficult for the worker to establish a satisfying psychological contract if he or she cannot come to an agreement about the work domain itself.[3]

Some workers feel that they have become invisible in an extremely ambiguous setting, and when they ask for recognition, they are expressing a deeply felt wish for validation of their identity. The following remark by a federal employee voices such a strong emotional need:

> I would like more feedback from my supervisor, probably. . . . I don't get to do anything in my name. I would like to get recognized as an entity. . . . It goes out in Bob's name; he gets called. But I have all the information, and it goes up and down the bureaucratic hierarchy. I like to think that my feeling about this is professional, rather than some personal identity problem, but who knows what evil lurks in the hearts of men and women?

Yet even here the worker's longing for recognition is tempered by her guilty ambivalence. Her strong sense of submersion in the organization is accompanied by the feeling that, perhaps, the treatment is warranted. Ambiguity about responsibility may be frustrating, but it may in some sense be structurally necessary. Perhaps—and here she introduces bad conscience—it is personally deserved.

Her undeniable suspicion that she may deserve not to be recognized is expressed in the successively more opprobrious interpretations she assigns to her concerns in the final sentence. Her wish for recognition may be properly professional, but she cannot be certain that it expresses anything more than a neurotic "personal identity problem." Still less can she be sure that it reflects anything more than the expression of "evil." Her inclusion of "women" in elaborating a common saying is evidence that she does seriously consider the possibility that her wishes are evil impulses for which she should feel guilty and be punished. Apparently she internalizes the structure of the organization and gives it social, and then moral, legitimacy.

Still, there are other workers who believe that they not only deserve recognition but actively encourage and organize it. If personal worth and identity are associated with recognition, then the greater the extent of one's recognition, the greater one's personal worth and the grander one's identity. A federal worker describes the effort to get credit for work done as "getting play in the press":

> Recognition: how much play you get in the press, so to speak, externally and internally. How much play do you get? Do people want to talk with you? Does the director have any verbal praise for what you have done? Is it effective?

Another worker expresses a common belief about the ways in which a reputation may be developed in the rumor mill:

> I am not atypical in looking at the standard kinds of things. It is nice to get a letter. I look for verbal feedback, formal or informal. I take great pride in the kinds of issues people get in touch with me on. For example, "We are thinking of automating everything; we want your advice." Versus "We are thinking of putting a desk calculator in Washington County." The degree to which my name is bandied around the rumor mill. I think I am a more exciting person as a result.

However, reliance on a rumor mill introduces an imaginary quality to recognition. The essence of a rumor mill is distortion of information. When responsibility is ambiguous, and consequently, recognition is threatened, a rumor mill may transmit and create information that contributes to a public identity. However, a rumor mill may also add to ambiguities about responsibilities and may contribute unpredictably to a distorted public identity. Workers who put stock in a rumor mill believe that they can emerge with a public identity which, if ambiguous, is nevertheless positive. Still, there is a possibility that reliance on the rumor mill encourages attention to appearance and reputation to the possible neglect of substantive work. The make-believe, or celebrity, quality of these public identities is suggested by the final sentence in the last quotation, in which the worker thinks he is "a more exciting person" as a result of having his name bandied about.

Thus there is a paradoxical quality to some public identities in bureaucratic organizations. On the one hand, they provide recognition, which is a remedy for the invisibility caused by ambiguous responsibilities. On the other hand, insofar as the public identity is at least partly a fantasy product of the rumor mill, it is not really personal recognition. For those workers who are concerned about being recognized for who they are, there may not be a fully satisfying response. Bureaucratic organization may be an inappropriate place to look for personal recognition.[4]

Generally, then, ambiguity about responsibility gives bureaucratic workers concerns about whether they will receive credit for their efforts. Many workers experience these concerns as threats to their identity as practitioners. In response, many attempt to create sufficient certainty about responsibility so as to give recognition to their efforts and themselves.

AMBIGUOUS AUTHORITY

Because these workers commonly experience ambiguity in both responsibility and authority simultaneously, their responses to ambiguity in responsibility are framed by their responses to ambiguity in authority as well. In referring only to responsibility, many fear that it may become ambiguously so small that they do not receive recognition for their efforts. However, with respect to responsibility and authority together, many bureaucratic workers, especially at lower hierarchical levels, are concerned that because the limits to responsibility are undefined, it may become ambiguously quite large while authority may become ambiguously quite small, with the result that they are recognized and held accountable for matters over which they have no control.[5] An apparent response to this combined ambiguity is not to seek to create certainty, but to create additional ambiguity.

The Peril of Recognition

An interesting story is told by a state worker who assiduously works to create a "celebrity" identity that will be recognized in the office rumor mill. In some detail she describes efforts to use feminine charms to develop a following among the men in the organization. She has changed her hair color, and she refers to a deliberate "personality change." Constantly, she says, "I check for rumors as to what people perceive of me." Nevertheless, she asserts, "I don't want to be recognized." What can she mean?

Apparently she means that the public identity that may be created in the rumor mill, for all the recognition it may bring, is not really a personal identity. Further, she does not want her personal identity recognized in the organization. She discusses her hesitation about recognition in the following way:

> I was working for X. He is a very political person. He is always currying favors of people in higher positions. The project we were working on was important. He invited me to meetings to impress me. I said, no, I didn't want to have dealings with the front office. *I don't want to be recognized.* People have been shot down so many times that they want to sit back at their desk and be left alone. You have a lot of people here who know their job, but they are not allowed to do their jobs.

This woman expresses common hesitations about establishing a certain identity within a sea of bureaucratic ambiguities. She suggests that acquiring recognition means getting out of one's (anonymous) place, an action that may bring reprisal from several quarters. Administrative superiors may regard her as insubordinate and, accordingly, shoot her down. If she were to succeed in getting recognition from her superiors, her peers would be jealous, because of a general belief that recognition is a scarce commodity. Once she were recognized, she would become a target for blame when problems occurred. Thus possible guilt about success and concern about others' resentment give rise to apprehension about being "shot down."[6]

Self-defense and Its Problems

Why ambiguous authority apparently encourages this ambivalence about recognition is suggested by a state worker's answer to a question about what he has authority to do: "Besides take care of myself?" In posing this half-joking rhetorical question, he actually says three things about authority in bureaucratic organizations. First, he feels that he has little authority and that others have significantly more. Second, he believes that others are likely to use their authority in some uncertain ways that will bring harm to him. Third, he is prepared to concentrate his efforts on taking care of, or defending, himself against others' potentially abusive uses of authority. He suggests that self-defense may take precedence over even the performance of formal organizational responsibilities. The first two beliefs support the view of the woman previously quoted that other workers, first, may be resentful of her recognition and, second, may have the ability to cause her harm in response. Accordingly, the belief in the primacy of self-defense appears justified.

When both responsibility and authority are ambiguous, ambiguity of authority presents this man with the risk that he may not be able to count on having sufficient authority both to carry out his assignments conclusively and to gain support for them. He risks being held culpable for any problems. Hence ambiguity of responsibility takes on a new meaning. His concern is not that responsibility may be ambiguously small, but that it may be ambiguously large. He may receive recognition, but the recognition may be disapproving. This fear is expressed by a city planner:

> Very infrequently do you have people tell you you did a good job. No news is good news. Then you have probably done a good job.

Self-defense would point to the strategy of reducing one's perceived responsibilities.

Yet such a direction is difficult to follow. Subordinates can hardly avoid their formal responsibilities. Further, as responsibilities are ambiguous, the actions required cannot be anticipated clearly. Nor, because supervisor's expectations are also often ambiguous, can the consequences be clearly foreseen. The execution of responsibilities carries a risk of offending a supervisor, whose authority may be exercised abusively.

This situation can make a worker anxious. Carrying out responsibilities while avoiding uncertain, feared punishments may require a delicate balancing act, as the following reminiscence from a senior federal civil servant suggests:

> The biggest ambiguity was the anomalous relation of a deputy to his principal. To me it became a matter of personal ethics. One thing I would never do is climb on my boss' back. So I always kept a certain distance from him, so that it could never be said that I was challenging him. . . . What I should have inscribed is that the boss one notch up should have delineated it. And I think I was penalized for *not* stepping into certain areas. To have done it would have made me something less than what I wanted to be. But I was not hungry to want his job. But if you had not taken it when it was vacant, they would say that you had not performed. I guess I should have, but it was not my nature. . . . They expected me to be a dynamo as a deputy when a dynamo would have threatened my boss. Should I get involved in challenging his status? It was not my responsibility to clear up those relationships unless I wanted to stir up a fuss.

This worker describes a conscious effort to carry sufficient responsibility so as to perform his job but not so much responsibility as to threaten his boss, whom he perceived as vulnerable. The ambivalence about recognition which such ambiguity stimulates is clearly expressed in his "but" statements. His final consideration evidently is his apprehension about the "fuss" that would follow inappropriate assertion on his part. His boss, though vulnerable, may respond to any perceived threat with punishment.

The subordinate does not know if his boss will exact punishments for specific offenses. Nor is the range of potential reprisals predictable, if they are forthcoming at all. Simply, at some point the subordinate's actions may be regarded as inappropriately aggressive, and the boss may somehow retaliate. Thus the subordinate risks acting in ways that unknowingly may lead to his being cut off by his boss.

Part of this ambiguity is objective. However, part is contributed by his ambivalent intentions toward his boss. The subordinate acts assertively for both conscious and unconscious reasons. Consciously, he does what he believes necessary in order to accomplish what his boss expects. Unconsciously, he may act aggressively against his boss—"stepping into certain areas," "challenging his status," even climbing on his superior's back—out of resentment of subordinacy, out of interest in advancing, out of interest in doing as much and as well as he can, or some combination of these reasons. Combined intentions both to challenge his boss and to avoid challenging him contribute to anxiety about punishment and guilt.

The anxiety experienced in contemplating action in the face of ambiguity is portrayed by a service worker whose responsibilities require him to handle many unique cases:

> There are no set structures. At any time I could be asked to do something that I have never been asked before. There are a lot of liability factors. I hope I would go to my supervisor and ask. . . . [Are unreasonable requests ever made?] It does not happen all

the time. There are several instances where I was asked to do something, and I said no. I could have not done it. However, I almost always do them. I feel it tends to make things run smoother. I have a good relationship with my supervisor. I consciously do this. I will let them know that [I disagree], either very strongly or indirectly. Indirectly, like in the form of a joke. If I am angry about something, I will show this in making a point. [If I am asked to do something unreasonable,] I will balk at that. I will ask them to direct me to do it. If and when they direct me to do it, I will do it. I have the option to take it to a personnel board. I will not do it. I view myself as more cooperative than the average city worker. But I think I am a little stronger than average when I am backed in a corner. Things that I consider against my judgment, my beliefs, I will not do it. . . . If [as punishment for not cooperating] there were a dirty job, I would get it. But if it is a requirement having to do with [the area of my primary job responsibilities, I would do it], sure.

He begins by identifying potential risks of working in situations where there are no structures. Then, repeatedly, he vacillates between asserting that he would do whatever he believes is necessary or proper and rationalizing obedience to his supervisor. He cites ethical principles and feelings of anger as indications of his independence, but simultaneously finds reasons or circumstances for going along with whatever his supervisor wishes. His response provides further insight into subordinates' responses to being controlled. When he is angry about something, when he is asked to do something unreasonable, he may momentarily balk, but he will then ask his supervisor to insist on the request, and he will comply as if he had no choice. To avoid the consequences of his own anger, he will organize the aggression of his supervisor against him, backing himself into a corner in which he is unlikely to harm others, who might then punish him.

This anxiety derives from the interplay of the organizational work situation and the worker's personal history. The following comments draw attention to the ways in which authority relations at work may evoke and play on childhood memories:

I have had two different supervisors. . . . They are two different individuals. . . . I suppose it is like your parents. You try to look objectively at each. There are good things about my father, and not so good things [and the same with my mother]. There are times when I look at them. There are always personal things. I suppose that when I was a child I developed a certain way. [Do you ever have overt conflicts with your supervisors?] No, an overt conflict . . . would be something that would aggravate me. I worry about things that I can do nothing about. There have been one or two overts. Once I went above the supervisor. It was the old back-into-the-corner routine.

In this example of transference, he identifies his supervisor with his parents and depicts himself as their child.[7] He worries, he says, about the one situation when he feels powerless—opposing his supervisor. The likely consequence of going against his supervisor, as, perhaps earlier, going against his parents, is being trapped (being told to sit?) in a corner. As just noted, he may even preemptively retreat into a corner in order to avoid confrontation. The phrase "old back-into-the-corner routine" tells two things about this experience. First, it is old, perhaps dating to childhood or infancy. In addition, the expression conceals any identification of who caused (or still causes) him to go into the corner—someone else, or himself.

Although not always conscious, these emotional themes, encouraged by the relationship between supervisor and subordinate, condition a subordinate's response to ambiguity about responsibility and authority. These themes frame a lesson for bureaucratic workers: even if self-defense should be an important concern, formal responsibilities are unavoidable, and acting to meet responsibilities risks placing subordinates in positions where, contrary to their intentions, they may be defenseless.

Counterambiguity: The Best Defense . . .

When ambiguous responsibility and authority make recognition apparently inherently risky, the main resort for subordinates is to attempt to reduce their perceived responsibilities and to avoid recognition. This might be done by negotiating formal agreements to reduce responsibilities to fewer tasks, but the ambiguities in tasks still carry risks. Alternatively, subordinates may attempt to create sufficient additional ambiguity about their responsibilities to confuse their supervisors and prevent the supervisors from punishing the workers for any perceived problems in meeting their tasks.

The basic form of this "counterambiguity" is "responsibility-without-responsibility," in which subordinates attempt to control definitions of responsibility so that they may withdraw from responsibility under cover of tactful obfuscation. A senior career bureaucrat presents the matter in this way:

> You are so adroit at covering yourself. Being vulnerable is something that bureaucracies are so concerned about. . . . I don't want to have freedom without getting anybody else's permission first. I have to use my own judgment. It is a very awesome responsibility, because it can be called back at any time. So that makes someone aware that it may be a risk to say that you have freedom.

Under these conditions it is possible to have responsibility only when another, a superior, has been persuaded to take responsibility for the actions involved. Such a negotiation may be attractive to a supervisor because it appears to offer some certainty about his own duties. Because task requirements are inevitably somewhat ambiguous, this co-optation of the supervisor into an agreement about responsibilities offers the subordinate the possibility of linking the supervisor to any problems that arise.

In doing this, a subordinate is supported by the tendencies of bureaucratic language, which emphasizes subordinates' lack of responsibility. Edelman (1977) explains,

> Language that emphasizes the subordinate position of the people who do the work further deflects criticism of their decisions. . . . This assertion . . . protects those who make the critical decisions from public criticism; it exalts superiors even when they are involved only rhetorically with the work for which they are credited; it degrades the status and the self-concept of subordinates even when they are making the critical decisions. (p. 96)

The essence of counterambiguity is "no-responsibility," in that the worker attempts to persuade supervisors that any responsibility is, somehow, someone else's. A state civil servant offers the following illustration, regarding her job description. In this case, rather than adding to ambiguity, the worker exploits an already existing ambiguity:

> I used to fight over job descriptions [when I worked] in the director's office. When I came here, a lot of employees were in jobs which no longer existed. I said, you need to change your description. They said, so what? They said, this gives an advantage to an employee. When you file for a job, if the personnel records are wrong, the only one who knows what you have done is you. You can't get locked in. To me everything had to be nice and neat. And what they showed me is that was not to their advantage. That is why I am enjoying this. Now I don't know what my job description says. Performance is evaluated in relation to the position description. I have not seen my position description, and I don't care. The only time is when I file for promotion. I don't want promotion. . . . If you know that there is a rule which says thus and such, you are responsible for carrying it out. If you don't know there is a rule, then you have no responsibility.

In the same vein, a federal policy analyst observes that colleagues enjoy exploiting niches of invisibility. Responsibility, he says, is ambiguous

> more often than not. Not whether you know what to do or not, but whether you have anything to do. A lot of people are envious because I have a study, something to do. But a lot of people are not looking for something to do. . . . Some people get upset; some are rather thrilled by it.

Menzies (1975) portrays a related process of creating—and institutionalizing—"responsibility-without-responsibility" among nurses she observed in a British general hospital. She argues that staff members unconsciously conspire to obscure responsibility through ambiguity in order to evade culpability for feared risks of complex duties:

> Additional protection from the impact of specific responsibility for specific tasks is given by the fact that the formal structure and role system fail to define fully enough who is responsible for what and to whom. . . . The content of roles and the boundaries of roles are very obscure, especially at senior levels. The responsibilities are more onerous at this level so that protection is felt as very necessary. Also the more complex roles and role-relationships make it easier to evade definition. . . . There is therefore a lack of stable person-role constellations, and it becomes very difficult to assign responsibility finally to a person, a role, or a person-role constellation. . . . Such diffused responsibility means, of course, that responsibility is not generally experienced specifically or seriously. (p. 294)

Further, in analyzing "the reduction of the impact of responsibility by delegation to superiors" (p. 294), Menzies shows that nurses attempt to extricate themselves from the burdensome responsibility of treating patients by assuming that only higher ranking nurses have the ability to perform major nursing tasks. Unfortunately, this re-allocation of assignments, in which the superiors tacitly collude, has a cost. In order to rationalize the shift, subordinates must split images of themselves and project their competence onto their superiors, while retaining feelings of competence for themselves alone. This process of splitting and projecting responsibility contributes to the recurrent feelings of incompetence and shame depicted in later chapters.

Additional insight into the motivations for counterambiguity is offered by the following testimony, which provides an example of what might be called "authority-without-authority." Here, a worker seeks a supervisor's collusion in acting as if a certain situation, with sufficiently unacceptable consequences, had virtually nothing to do with the worker.

> When you have some latitude, when the managers over you trust you, you have some authority, but you have to do it in accordance with the system. . . . As long as I am bringing credibility on my agency, I have the authority to do anything. If I bring discredit, I don't have the authority to do anything: all of it is delegated. I rather enjoy that. There are a lot of advantages of being in that kind of mode. One can have various escape hatches.

Although immediately concerned with authority, rather than responsibility, this approach permits the worker to sidestep any action which, if it had occurred, would have carried both responsibility and culpability. A supervisor may be encouraged to collude in such an arrangement by a desire to erase the unfortunate consequences of an action. Thus only creditable, successful actions have consequences or reality. If any problems occur, the subordinate never had the authority to do what he or she did, and unauthorized actions carry no responsibility.

"Authority-without-authority" reveals the creation of counterambiguity to be the bureaucratic social expression of the psychic defense mechanism of undoing: regrettable actions are reversed. As later chapters will show, undoing is the worker's typical response to the predicaments and problems of bureaucratic work. As such, undoing has two important meanings for workers. First, it defends them against anxiety by removing actions that could bring punishment. Second, it represents an act of revenge against the organization that causes the anxiety. By removing problematic situations from the organization, it undoes the organization, preventing anxiety-inducing situations from arising in the future and crippling the organization from carrying out its mission.[8]

Counterambiguity bears several potential problems. It represents an effort to cause a supervisor to see a subordinate's responsibilities as more ambiguous than the supervisor might have considered them. If counterambiguity succeeds, a subordinate may thwart a supervisor who appears threatening from using evaluative authority in a potentially abusive way. The primary limitation on counterambiguity is that a supervisor must at least tacitly agree to see the situation in the more ambiguous way that the subordinate suggests. However, supervisors, as any bureaucratic workers, require some indication about who is responsible for what, and, in particular, who other than themselves may be held culpable for any problems. Ultimately superiors have the authority, whether their perceptions are accurate or not, to enforce any definition of responsibility on their subordinates. Moreover, in the fog of counterambiguity created by a subordinate, a supervisor may even punish the subordinate for problems that a less amorphous view would have associated with another worker. Thus there is no certainty that counterambiguity will succeed.

On the other hand, there may be a problem if counterambiguity does succeed. Bureaucratic workers may effectively cast themselves into positions in which they have neither authority nor responsibility. Unfortunately, although this position provides an escape from feared abuse by superiors, it offers the extricated workers no sense of competence or identity. Worse, it is a self-inflicted invisibility. Recognition of the no-win character of counterambiguity may refuel efforts to undo the organization.

Thus the combined ambiguity of responsibility and authority encourages an effort to escape responsibility. One feasible strategy for doing this is to create sufficient additional ambiguity that personal responsibility becomes obscured. And yet such an action, whatever its immediate consequences, seems certain to defeat workers' continuing interests in gaining recognition for their efforts.

AMBIGUITY AND COUNTERAMBIGUITY: THE BUREAUCRATIC PREDICAMENT AND PROBLEM SOLVING

Some ambiguity is inherent in tasks, and some resides in the social structure of bureaucratic organizations created to accomplish the tasks.[9] What these accounts illustrate is how normal unconscious psychological processes are provoked and channeled by specific bureaucratic ambiguities in responsibility and authority, creating an ever more complex situation. These processes build up a psychological structure that complicates problem solving.

Many bureaucratic workers respond ambivalently to organizational ambiguity. On the one hand, ambiguous responsibility leads them to try to make their responsibility clear. On the other hand, ambiguous authority leads them to create counter-ambiguity to avoid responsibility. In confronting organizational ambiguity, then, bureaucratic workers may simultaneously feel needs to both clarify and confound the situation. Because feelings about personal identity are at stake, both impulses may be quite strong. Anxiety about the possible consequences of any decision intensifies the conflict. This ambivalence has the potential to immobilize workers by creating indecisiveness about which direction they want to move in. As much of the ambivalence is unconscious, workers may not be able to clarify their intentions about a situation. All of these results hinder competent work, creating more frustration.

One common way in which people may resolve ambivalence is to split it and project one element onto others. Thus bureaucratic workers might embrace the more acceptable impulse (seeking to clarify an ambiguous situation in order to acquire responsibility) and project the less flattering one (making the situation more confusing) onto their supervisors. Then it is apparently the supervisor who makes situations ambiguous. Menzies' (1975) account of the nursing service provides an example. The nurses (consciously) complained about conditions which they (unconsciously) created.

In this case, a subordinate needs to work still harder to force some clarity into a situation. With the clarifying and confounding impulses in this way apparently separated, a worker can see him- or herself striving all the more forcefully to clarify a situation that a supervisor, perhaps with the intention of confusing and then abusing the subordinate, aims to render still more ambiguous. Although this act of projection defends the subordinate from responsibility for tasks where blame might be assigned, it also prevents task definition.[10]

One consequence is that workers cannot distinguish between real and imagined ambiguities. Consequently, they have difficulty selecting and executing actions instrumental for solving problems. Yet the bureaucratic psychological structure still more seriously interferes with problem solving by curtailing the will to work on problems. The predicament of exercising responsibility without authority encourages workers to retreat from recognition out of fear of the potentially abusive character of power. While seeking to avoid envy or scapegoating, workers are unlikely to enter into collaboration with others. To the contrary, they are likely to frustrate others'—and ultimately their own—efforts to exercise power in attacking problems.

This chapter indicates why any resulting anxiety about bureaucratic authority may not simply lead to retreatism or apathy. At the same time that the predicament creates apprehensions about the costs of assertion, it holds out a continuing threat to personal identity which only acts of assertion can remedy. This predicament continues to motivate bureaucratic action.

NOTES

1. As noted already, responsibility and authority may be both formally designated and informally constructed. Formally, responsibility normally exceeds authority. As this book reveals, workers' unconscious assumptions about responsibility and authority tend to increase the experienced inequality. Tannenbaum et al. (1974) report that studies consistently find bureaucratic workers complaining about the inadequacy of authority for responsibility: "Results from a number of studies show that workers, supervisors, and managers are more likely to feel that they have too little authority than too much. . . . Furthermore, this felt deficiency decreases with hierarchical ascent" (p. 7).

2. For an explanation, see Baum (1980, 1983a). It should be noted, however, that position in the organizational hierarchy influences expectations. Among planners, for example, while staff subordinates are primarily concerned about recognition from colleagues, administrators are more concerned about concrete results and recognition from political clients.

3. Sennett and Cobb's (1972) study of industrial workers found that efforts to divorce personal identity from work for the organization fail more often than not.

4. Sociologists say essentially the same thing when they observe that formal organizations are constituted of roles, not persons. For example, see Berger (1963).

5. As Tannenbaum et al. (1974) reported, experienced inequality between responsibility and authority is expectably greater at lower hierarchical levels. However, virtually all bureaucratic workers are subordinate to someone; even the chief administrator is accountable to someone outside the organization. Although workers at the bottom perceive the greatest difference between responsibility and authority, this discrepancy is common throughout the system and contributes to the creation of additional ambiguity.

6. For further illustration and analysis of how such feelings and fears contribute to ambivalence about recognition, see Cohen and Cohen (1951) and Gustafson and Cooper (1979).

7. Clearly, although he alludes here to ways in which his supervisors remind him of his parents, additional discussion and analysis would be necessary to identify specific perceived or felt similarities. Examples of the range of parental feelings aroused by supervisors are offered by Hodgson, Levinson, and Zaleznik (1965), and Kets de Vries and Miller (1984).

8. For an analysis of undoing as a defense mechanism, see Fenichel (1945), Freud (1946), and Freud (1936; rep. 1963).

9. How much of what is ambiguous in organizations is always an empirical question. Organizations vary along a continuum of unclear to certain. Moreover, the same organization may be more and less ambiguous at different times, or some parts of an organization may be

more amorphously defined than others. Responsibilities may be clearly defined, while authority may not be. Or some responsibilities may be certain while others are ambiguous. For a thoughtful summary of these questions, see Perrow's (1977) review of March and Olsen's (1976) book on organizational ambiguity.

10. For a discussion of projection as a defense mechanism, see Fenichel (1945), Freud (1946), and Riviere (1937; rep. 1964).

4

Autonomy, Shame, and Doubt

You never win all the contests and it is unpleasant to lose for whatever reason. But when you win because of your professional skills alone or even in combination with your political abilities, the victory is sweeter. When the *only* thing that stops the street from getting widened is your ability to "call out the troops," that is, your political skills, that is not a very good victory. It is a little bit shameful. The planning commission, the mayor, or the city council could as well hire anyone else; they don't need a planner. (Allan Jacobs, *Making City Planning Work*)

"Please wait a moment" [K. asks as the priest is about to leave]. "I am waiting," said the priest. "Don't you want anything more from me?" asked K. "No," said the priest. "You were so friendly to me for a time," said K., "and explained so much to me, and now you let me go as if you cared nothing for me." "But you have to leave now," said the priest. "Well, yes," said K., "you must see that I can't help it." "You must first see who I am," said the priest. "You are the prison chaplain," said K., groping his way nearer to the priest again; his immediate return to [work at] the bank was not so necessary as he had made out, he could quite well stay longer. "That means I belong to the Court," said the priest. "So why should I want anything from you? The Court wants nothing from you. It receives you when you come and it dismisses you when you go." (Franz Kafka, *The Trial*)

Why does Allan Jacobs, former director of the San Francisco Planning Department, say that he felt ashamed when he used political power to win a decision? What is preferable about the power of technical analysis? Why does he worry about being released when he acts politically? Whereas his technical skills distinguish him from others, political skills do not differentiate him from anyone else, and he could just as well be replaced. Dispensability is the essence of shame and is a continual worry in a hierarchy, as Kafka's more dramatic example shows. When superiors are indifferent to their subordinates' fates, subordinates tend to feel worthless. K. reveals the results: they refrain from asserting themselves or arguing their positions; instead, they accept their supervisors' apparent devaluation of them. Finally, just as K. no longer feels it necessary to return to his job, people who are ashamed lose the drive to continue with their work. This chapter examines the ways in which bureaucratic authority may discourage many workers, including top administrators, from acting powerfully, even when reasonable positions call for such initiative.

Collaborative problem solving requires a willingness to take responsibility, as well as security exercising power. However, as observed earlier, many workers are confused about whether and how bureaucratic power is exercised, and they are ambivalent about wielding it themselves. This chapter focuses on planners to look directly at subordinates' relationships with those who have significant organizational authority. It is not simply the disparity in authority between superior and subordinate but also the peculiar way that subordinates experience bureaucratic authority that deter them from acting and confuse them about the reasons for their immobility. The workers in Chapters 2 and 3 think in terms of guilt and anxiety about retribution for aggression toward authority figures. This chapter shows how bureaucratic authority may give rise not only to guilt but also to shame, and how the two are related.

As other bureaucrats, planners are ambivalent about exercising power. They simultaneously say that they want power and yet express distaste for exercising power (Baum, 1983a, 1983b; Clavel, Forester, & Goldsmith, 1980; Vasu, 1979). For example, a study of planners' espoused ethics finds that more than two-thirds advocate courses of action considered highly political (Howe & Kaufman, 1979). Nevertheless, other studies indicate that most planners regard themselves as primarily intellectual practitioners (Baum, 1983a, Needleman & Needleman, 1974), many feel themselves relatively powerless (Baum, 1983a), and many resist any explicit efforts to exercise power (Baum, 1983a; Cole, 1975; Vasu, 1979). This ambivalence clearly interferes with the planners' expressed interest in exercising power to solve problems.

One explanation for this ambivalence toward power could be found in the psychological traits of people who choose to practice planning. There is some indication that many who become planners wish to influence others without having contact with them and anticipate that their activities will control others from a secure distance. As a consequence, any overt exercise of power makes them uneasy (Baum, 1983a). But in addition, there is something about the nature of bureaucratic authority that, regardless of members' personalities, appears to contribute to this ambivalence. A common form of bureaucratic authority, autonomous authority, may evoke feelings of shame and doubt that discourage many bureaucratic subordinates from exercising power. Before examining this phenomenon further, it is useful to look more closely at what "power" means.

THE MEANING OF POWER

Generally, power may be understood as an interest in feeling strong or having an impact on others. However, it is a troublesome concept and experience, and is rarely discussed explicitly. Planning practitioners, for example, commonly assume that their activity does not involve the exercise of power, or at least does not reflect any intention to be powerful. Sometimes planning is presented as an activity in which planners organize information in order to enable the information to be powerful, while planners remain, at most, only faithful intellectual servants. In contrast, "politics" is the label given to all activities in which people attempt to be powerful. And yet, with few exceptions, people become planners in order to influence decision making, which unquestionably indicates their interest in power. Thus the common distinction between planning—or administration—and "politics" is not only inaccurate, but obfuscates issues of power and makes it difficult for planners to understand how power enters into their work.[1]

Planners' ambivalence about power is not unique. The effort to deny any inclination toward power is common in contemporary Western culture because of the ways in which power is normally conceived. For example, although power may be exercised for either constructive or destructive purposes, in our culture we tend to assume that power may be exercised only destructively. The possibility of irreversible, ultimate, and total devastation of the world by nuclear power may effectively symbolize the danger in any exercise of power.[2] In addition, although power may be directed either toward strengthening oneself or influencing others, in our culture we tend to assume that it is only directed toward others, and, because we also assume that power is only destructive, we tend to deny the presence of power and any interest in or inclination toward it. This makes it difficult to consider its constructive uses (May, 1972; McClelland, 1975). This negative image may lead those who have to exert power to feel guilty. The peculiarly American experience of individualism gives this image of power as the initiation of destructive conflict with others an especially sinister tone. The aggressive impulses of power are conceived as an individual challenge of a multitude, with one possible outcome the destruction of the lone individual. To any guilt about aggression is added anxiety about its anticipated end.

This cultural image of power as aggressive and destructive may create an ambivalence about exerting influence. In addition, these conflicts are accompanied—and may be encouraged—by unconscious psychological conflicts. Moreover, a wish to avoid the specific emotions aroused—anxiety and guilt—would also lead to a denial of power.

A Framework for Thinking About Power

Unfortunately, the confusion about power, and its consequent denial, obscures the ways in which planning inherently involves power. The actions of planners involve exerting power in specific directions, for example, in selecting which problems receive attention and what alternatives or solutions are to be implemented (Forester, 1981; Seeley, 1967). Additionally, as noted before, choosing to become a planner implies an intention to influence social decisions. Further, there are numerous ways in which planners continually express specific powerful intentions.

These aims can be understood in a framework developed by McClelland (1975) for analyzing intentions to be powerful. He suggests that a need for power includes any concern about having an impact on action, feeling, or values. His framework involves both constructive and destructive power, both self-affecting and other-affecting power. He differentiates power orientations along two dimensions: the source of power (oneself or another) and the intended object of power (to strengthen oneself or to influence another). Each of the four identified orientations has a unique intention and a distinctive expression in action. Table 2, adapted from McClelland's work, shows these different intentions.

With this framework it is possible to characterize the activities involved in planning according to their power orientations. Most planning activities fall into the second or fourth category; thus planning is described as a sequence of activities in the second and fourth orientations. Type II activities are concerned with becoming strong and powerful by increasing the accumulation of information and objects and by regulating their release. This category encompasses the data collection activities

Table 2. Classification of Power Orientations

Object of Power	Source of Power			
	Other			Self
Self (to feel stronger)				
Intention:	"It" (God, my mother, my leader, food) will strengthen me.			I will strengthen, control, direct myself.
Action:	Being near a source of strength.			Collecting, accumulating information, things.
		Type I	Type II	
Other (to influence)		Type IV	Type III	
Intention:	It (religion, laws, my group) will move me to serve, influence others, to do my duty.			I will have an impact (influence) on others.
Action:	Action on higher principle or purpose.			Competing with, affecting others.

Adapted from McClelland (1975), p. 14.

of planning, along with planners' concerns about who will receive the data once analyzed. Type IV activities are concerned with becoming strong and powerful by serving some higher principle or purpose. This category includes the activities in which planners make recommendations on the basis of an interpretation of the public interest or technical or economic rationality. Thus one common version of the planning process may be seen as a sequence of Type II power-oriented collection of data, Type IV power-oriented interpretation of data and formulation of recommendations in the light of rationality and the public interest, and Type II power-oriented release of the interpreted data and the recommendations. These tasks comprise the conventional description of the planning process—in particular, the process distinguished from the "politics" of power.

McClelland's framework indicates that these routine planning activities do express intentions to be powerful. Significantly, the framework helps clarify the distinction between Type II and Type IV power orientations in planning activities and what is commonly implied in references to "politics" and power. Types II and IV orientations differ from Types I and III in a crucial respect. Types II and IV both have an identical source and object of power, whereas Types I and III have different sources and objects. Type II has the self as both source and object, so that the actor may empower him- or herself. Type IV has an other as both source and object, so that the actor may be the representative of an abstract principle or group identity. This relationship may be called suprapersonal. Although one person may act vis-à-vis others, and although personal identity may be at risk, the stakes involve abstract principles or group interests, rather than self-interests. Neither the Type II nor Type IV orientation aspires to an interpersonal relationship in which one's personal interests are directly set before others. This is the essential difference between the power orientations involved in activities identified with planning and those identified with "politics."

This difference may be clarified by contrast with examples of planning activities that would represent either a Type I or Type III power orientation. The Type I orientation, with an other as a source of power and the self as the object, would include, for example, acting as an advisor to a political leader. The Type III orientation, with the self as a source of power and an other as the object, would include, for example, advocating or lobbying for recommendations with administrative superiors, colleagues, or elected officials. The personal confrontation in these roles is most consistent with conventional Western definitions of "power," and it is the focus of planners' attention when they discuss "politics" and debate the propriety of acting "politically."

The crucial point illuminated by McClelland's framework is that the decision to become engaged in "politics" is not a decision to become involved with power, but rather a decision to attempt to acquire power in particular ways. The focal issue for planners who already express combinations of Type II and Type IV power orientations is whether to consider a Type III power orientation, which would thrust them into competition and conflict with other people in directly interpersonal relationships.

AUTHORITY IN BUREAUCRACY

Thus the Type III orientation expresses the kind of power about which many planners and other bureaucrats have ambivalence. In order to understand why practitioners may retreat from a Type III orientation, despite its potential rewards, it is necessary to analyze bureaucratic authority. First, authority is expressed in most bureaucratic organizations in such a way that the exercise of power is barely visible. Second, authority in bureaucratic organizations is frequently accompanied by feelings of shame and doubt that unconsciously inhibit workers from moving from a Type II power orientation to a Type III power orientation.

The Double Bind of Bureaucratic Power

Bureaucracy is built on a double bind concerning Type III power. On the one hand, the primary, Weberian, rationale for bureaucracy is "the elimination of power relationships and personal dependencies—to administer things instead of governing men" (Crozier, 1964, p. 107). Thus the ideal bureaucracy would limit the assertion of personal power, and the model organizational member would not attempt to exert it. The corresponding message communicated by bureaucratic norms is that proper behavior for membership precludes the exercise of personal power to attain goals. However, in real bureaucratic life power proves necessary both to resist harmful consequences of others' actions and to satisfy one's needs—for example, to do competent work. Thus the conflicting message communicated by bureaucratic practice is that attainment of goals within the organization requires powerful action, which is improper and implicitly a ground for exclusion from the organization. But apparently effective work and the maintenance of meaningful personal identity require acting powerfully within the double bind concerning the proper place of power in bureaucracy (Zaleznik, Kets de Vries, & Howard, 1977).

The double bind impedes powerful action by presenting contradictory directives. No sooner may workers think about what they need to do to be effective, than conflict about the actions required for effectiveness confuses their thinking. They may fear that if they attempt to act effectively, they will be expelled from the organization. Nor is it easy to analyze others' exercise of power and calculate reasonable strategic responses, for mixed messages disagree about whether power is really exercised. The mixed messages thus conceal the double bind and further hinder powerful action.[3] Moreover, the feelings associated with power make it difficult to confront and resolve this confusion.

The Autonomy of Bureaucratic Authority

People in society exercise varying amounts of power and control over one another. Normally, people consciously attempt to interpret these relationships in ways consistent with personal dignity and so giving the relationships an acceptable rationale. Hence, as Sennett (1980) observes, power to control others is commonly translated into images of strength, or authority. Those who have power over others are assumed to have superior strength of some valued kind. In this respect, Sennett argues that all authority contains elements of what Weber (1964) distinguishes as "charismatic" authority: the endowment of leaders with "supernatural, superhuman, or at least specifically exceptional powers or qualities" (p. 358). As this chapter will show, real bureaucratic authority, in contrast with its ideal type, ambiguously combines charisma with legality.

The conscious rationalization of power as strength is in accordance with unconscious memories from infancy: the parents on whom one is dependent are certainly much stronger, if not omnipotent. Transference of this early belief onto adult contemporaries reinforces the overt view that control is strength—that it is somehow irresistible and perhaps also good. Freud (1921; rep. 1959) has described how members of a group normally attribute this extraordinary strength to their leader. Unconsciously, members collectively tend to endow their leader with the qualities they earlier identified with their parents. Thus a group leader, as other social leaders, may be regarded as an "ego ideal": group members follow their leader because they unconsciously identify this person with all their infantile ideals regarding achievement, mastery, power, and independence.[4]

Adults need to follow authority for the same reasons that infants depend on it. Authority that is good not only controls, but also guides and reassures. However, authority may impose discipline, and bad authority may inspire fear. Hence it is possible, as Sennett points out, to perceive someone as authority—as strong—and yet question the legitimacy or moral value of the undeniably superior strength.[5] This view of authority differs from that of Weber (1964), who defines authority as legitimate power only. In both views, authority represents an interpretation of social power that is in some way acceptable to its subjects. Sennett, unlike Weber, gives particular weight to unconscious needs to respond to authority, even in conflict with conscious beliefs that authority is unjustly or illegitimately exercised. One unfortunate consequence of this process of interpreting power, perhaps especially evident in bureaucracy, is that

> The dilemma of authority in our time, the peculiar fear it inspires, is that *we feel attracted to strong figures we do not believe to be legitimate.* (Sennett, 1980, p. 26)

The importance of such ambivalence becomes clear later in this chapter, in connection with anger toward hierarchical authority.

Bureaucratic workers commonly see organizational authority in ways that incorporate the images of strength just described. Sennett has selected a helpful term to characterize this authority: it is "autonomous," or self-sufficient. People with autonomy have expertise and traits that designate their ability to satisfy their own needs independently, without others. Indeed, the autonomous authority is indifferent to others.[6]

The administrative literature offers frequent portrayals of autonomous bureaucratic authority. Henry (1949), for example, reports of the business executive that

> he looks to his subordinates in a detached and impersonal way, seeing them as "doers of work" rather than as people. He treats them impersonally, with no real feeling of being akin to them or of having deep interest in them as persons. (1949, p. 290)

Crozier (1964) observes that, "Generally speaking, workers have an image of their director as cold and distant and having no consideration for the status aspirations of his workers" (p. 82). The autonomous bureaucratic administrator has neither need for nor interest in the subordinate.

The indifference of someone who is powerful simultaneously poses a riddle and increases that person's control. The riddle concerns the identity of someone who exercises such undeniable authority in virtual anonymity. How does this person do it? And what does he or she want from subordinates? Efforts to solve the riddle lead to greater attentiveness to and, consequently, dependence on the authority. Sennett describes this process:

> [The] very acts of indifference maintain dominance . . . In reacting to this dominance, those in need can come to perceive autonomous figures as authority . . . Someone who is indifferent arouses our desire to be recognized. . . . Afraid of his indifference, not understanding what it is which keeps him aloof, we come to be emotionally dependent. (1980, p. 86)

Thus the indifference of autonomous authority intensifies the wish for recognition already evoked by the ambiguity of bureaucratic responsibility and attracts subordinates to persons with such authority.

However, because autonomous authority figures are apparently indifferent, subordinates must act "for" them. As described earlier, subordinates tend to fill in characteristics of bureaucratic authority by transferring to them cognitive and emotional assumptions associated with early life experiences with authority. These childhood experiences contribute an imposing quality to impressions of bureaucratic authority. On the basis of these imputed characteristics, subordinates tend to imagine what the authority would want of them and then to internalize these demands and enforce them themselves. This process, called "introjection," provides some specificity regarding the authority's expectations. In addition, it helps to defend against the anxiety the authority arouses by taking some apparent personal control over the authority's actions.[7] The resultant images of the imagined wishes of the authority serve as a guide for action. However, because these processes are unconscious, it is difficult to identify ways in which perceptions or feelings about authority may be inappropriate to actual bureaucratic situations and to develop accurate perceptions of the authority relationship.

EXAMPLES FROM PLANNING

This conceptualization of autonomous authority helps explain planners' descriptions of bureaucratic authority. Structurally, both public planning agencies and private planning firms have shallow hierarchies, with the consequence that most staff members are in positions of being subordinate to others without themselves being superior to anyone. The minority position of those who do occupy administrative or supervisory roles reinforces the superordinate-subordinate relationship. The following account is a composite of examples offered by many, though not all, planners interviewed. The situations and experiences are most common in larger bureaucracies, such as major city or county planning agencies, in which hierarchic relationships are more formal.

A planner's work normally begins when an administrator issues an assignment.[8] The assignment may be extensively detailed, but it may also be perfunctory, so that the administrator simply requests an analysis or a report on an issue. The assignment may not be presented in a face-to-face meeting, but, instead, may be transmitted in a written memorandum. Instructions often are sketchy, for several reasons, such as ambiguities in what the task requires, how it could serve organizational goals, and how it fits conventional programs (Lipsky, 1980). The administrator may assume that a competent planner should be able to figure out what the assignment requires and do it appropriately. At the same time, a planner may interpret the brevity of the request as an indication that the administrator regards responsibilities for other staff members as too important to permit detailed discussion of this assignment. While the planner works on the project, he or she may have little contact with the administrator, and consequently, may receive little or no indication whether the approach taken is acceptable. In addition, the planner may receive no hint as to the administrator's assessment of the quality of the work. The administrator may be occupied with numerous other matters about which the subordinate knows nothing. For example, a planner preparing an analysis in one area may be unaware that a director is under pressure from elected officials for the delivery of a report in another area. On occasions when the administrative supervisor passes through the space where the subordinate works, the administrator may be preoccupied with other matters and may take little notice of the subordinate and the work the administrator assigned.

These conditions make it difficult for planners to assess their work. A county staff member notes that he can not evaluate his work "when I have not been able to participate in policy making that is asking to have that particular item done." A state analyst observes that because tasks vary, not even her supervisor can clarify expectations:

> Part of [my difficulty assessing my work] is that I am new on the job, and because of its nature that it is not a routine job, I am not sure what the expectations are on any

assignment. Sometimes [my supervisor] doesn't know either. Because things are not routine, next week I may get an assignment that I have not done in six months. It is constant on-the-job training in this situation. And sometimes, because of the way some of the assignments come through, we are not really sure what somebody else's goal is for the assignment.

Nevertheless, as the planner understands, the administrator will judge the planner's work. Therefore, facing an administrator who may give the impression of finding the worker dispensable, a planner may anxiously watch the administrator's gestures and words for some indication of approval. A planner may listen for covert messages in statements in staff meetings and may assiduously work the office rumor mill for hints of the administrator's opinions. In considering their effectiveness on projects, subordinates may almost unconsciously discard any concerns about implementation or adoption of recommendations and focus directly on getting compliments or praise from administrators.[9]

Answers to a question about what they look for as indicators of their effectiveness, reveal planners' emphasis on their wish to be recognized in a superior's evaluation. A housing planner says his primary concern is "being taken seriously by my director." A land use planner says

As far as superiors go, you kind of look for indications that they are giving your views some considerations, whether they are willing to set up meetings to hassle down something.

The rarity of these responses from supervisors gives them increased value, and until the subordinate receives an evaluation, he or she may spend considerable time imagining and attempting to respond to the administrator's possible criticism. A county education planner tries to create the responses that do not come from top administrators. In evaluating his effectiveness, he seeks

responses from other people, verbal responses, how people react. When I do workshops, I give out feedback sheets. I've even started doing this with our staff after our Friday management meetings. I haven't perfected this yet, but yesterday I had good results.

Another planner, quoted earlier, vents a common complaint:

Very infrequently do you have people tell you you did a good job. No news is good news. Then you have probably done a good job.

The zeal motivating such concern for administrative approval reflects the planner's concern about not only evaluation of the project but also confirmation of his or her competence. In this process—where the administrator may communicate relatively little and the planner may continually look for approval from above—the administrator acquires increased authority.

Crozier (1964) reports that workers in French bureaucracies react similarly:

Directors command . . . a great deal of attention, and the reactions of their subordinates to their personality and behavior indicate another aspect of the formal authority relationships. This aspect concerns managerial prestige and the indirect influence directors may acquire because of this prestige. . . . [A chronic complaint among subordinates is] the problem of contacts. Employees consistently complain about the existence of a gap between them and the directors. They feel that no communication is possible. The

> directors cannot understand, and there is no way to have them understand, *what it is really like.* . . . Workers seem to be extraordinarily sensitive to the formal care and attention they receive from him.
>
> > He never says good morning. When talking he will keep his hands in his pockets. You cannot talk to him, he does not pay attention. (1964, p. 82)

Frequently, final decisions about agency positions on issues on which the planner has worked are made privately among agency administrators, out of the subordinates' view. No explanation may be offered for decisions, including those in which a particular planner's work is not fully accepted. If an explanation is given, it may cite general agency goals or concerns with no specific connection to a staff member's assignment. An administrator, in the explanation, may refer to information about external threats to or opportunities for the agency, which the specialized subordinate is unlikely to have or be able to challenge. A federal housing analyst concludes that she can never be certain what her supervisor makes of her work: "I like to think that she respects the work I do, but I don't know." These unresolved uncertainties add to the apparent arbitrariness of administrative action.

These examples illustrate the double bind described earlier. The relationship between a planner and an administrator is not simply one in which two persons reason together and seek to persuade one another about the most rational approach to issues. The administrator has control over whether the planner's recommendations are accepted and which conclusions are endorsed. At the same time, the exercise of this authority is virtually invisible, which confuses the issue of whether power is exercised.

In addition, the subordinate is unsure about whether he or she should attempt to use power—either alone or in concert with others—to gain the administrator's attention or to influence the administrator's judgment about the quality or use of work. If the administrator has not exercised power, then no powerful response is called for, and none is appropriate. If a subordinate concludes that an administrator, indeed, is exercising power, problems arise if the subordinate attempts to respond to the administrator's power in kind. Any attempt to imitate an administrator's autonomy assumes a self-sufficiency and independence unrealistic for subordinates; subordinates who attempt to control their availability to administrators would not fit into work routines and would eventually be punished for insubordination. A subordinate may develop other strategies for influencing a superior, but the basic hierarchical asymmetry of dependencies limits the subordinate's efficacy and impunity.[10]

Interpretation: Reality and Transference

Is this composite account of autonomous authority in planning organizations an accurate record of administrators' actions? There are reasons to believe that planners may exaggerate the autonomy of their supervisors. The shallow hierarchy of most planning organizations tends to highlight the superior-subordinate relations. In addition, planners as a group tend to have a relatively poor understanding of the complexities of organizational relationships. Many simply do not recognize differences of interest, and they are not interested in decision-making processes that result from and permit expression of conflicts of interest. Rather, most tend to conceptualize their work as the organization of ideas and information, which should require a linear process of increasing clarification and agreement. The cognitive maps of most planners depict little detail regarding organizational position, turf, interests, or processes of decision making (Baum, 1982, 1983a). Consequently, planners are particularly likely to conceive of all organizational actors who may constrain the progress of their intellectual products as homogeneous administrative "others," thus reinforcing the dualistic view of administrators and subordinates.[11]

Nevertheless, there is some indication that many administrators act in autonomous ways according to planners' descriptions. Although no comprehensive inventory of administrator styles is available, strong suggestive evidence comes from Argyris and Schön's (1974, 1978; Argyris, 1982) examination of their implicit intentions. Argyris and Schön report that one set of action strategies is dominant among the subjects in their sample, which includes planning administrators:

1. *Design and manage the environment unilaterally* (be persuasive, appeal to larger goals).

2. *Own and control the task* (claim ownership of the task, be guardian of definition and execution of task).

3. *Unilaterally protect yourself* (speak with inferred categories accompanied by little or no directly observable behavior, be blind to impact on others and to the incongruity between rhetoric and behavior, reduce incongruity by defensive actions such as blaming, stereotyping, suppressing feelings, intellectualizing).

4. *Unilaterally protect others from being hurt* (withhold information, create rules to censor information and behavior, hold private meetings). (1974, pp. 68–69)

This is a clear description of the exercise of autonomous authority. A dominant theme is concealment of evidence from others, speaking "with inferred categories accompanied by little or no directly observable behavior," and avoiding confrontation over mutually observable data. Argyris and Schön describe the consistent effects of these action strategies on others in terms virtually identical to those postulated for autonomous authority. For example, when actors seek unilaterally to protect themselves from confrontation, "the effect, intended or unintended, is to force others to guess at their meaning" (1974, p. 71), perhaps by transferring meanings from apparently analogous situations. Although Argyris and Schön have not systematically surveyed administrators or organizations, they conclude from their own and other studies that these action strategies dominate all or most organizations (Argyris & Schön, 1978). People find these strategies attractive because, by projecting the image of self-sufficiency, they defend against personal vulnerability. Actors with power could be expected to employ these strategies to protect themselves at the expense of others.[12] This defensive posture may be particularly desirable in planning agencies, which are the targets of conflicting public pressures to solve complicated and controversial problems.

One might argue that planners' peculiar misunderstandings of organizations encourage inaccurate perceptions of administrators, including the formation of images from transferred ideas. However, evidence about administrators' action strategies suggests that there is some reality to planners' perceptions, and that the administrators themselves encourage this transference of thoughts and feelings. Both processes occur within pyramidal organizations, which tend to limit direct contact between administrators and subordinates, encouraging transference responses and, thus, idealization of authority figures.[13] The complexity of bureaucratic relationships makes it difficult to separate those parts of planners' images which are based on transference from those founded in reality. Furthermore, administrators who are the objects of transference may unconsciously accept the transferred images as role expectations and enact them. Thus subordinates' transferred assumptions, while initially unrealistic, may gradually become valid observations.[14]

HOW AUTONOMOUS AUTHORITY IMPEDES LEARNING TO ACT POWERFULLY

A Framework

The framework suggested by McClelland helps explain why bureaucratic workers may have difficulty initiating powerful actions in response to autonomous authority. In that framework, autonomous authority and any powerful action against it would be considered expressions of a Type III power orientation, or interest in interpersonal power, in which one has some impact or influence on another. This orientation contrasts with Type II, which is an interest in intrapersonal power, and Type IV, which is an interest in suprapersonal power.[15]

One important characteristic of McClelland's framework has not yet been mentioned. McClelland regards the four types of power orientation as four stages in a sequence of personal development. He contends that it is necessary for someone to master a Type I orientation before learning a Type II orientation, that mastery of Type II is necessary for learning Type III, and so forth. Once someone has mastered a particular stage, then this person is able to act in that orientation or any preceding orientation while attempting to master the next one. McClelland formulated the four stages on the basis of empirical examination of voluminous descriptions and images of power in many cultures. He concluded that these distinctive expressions correspond to the four stages of psychosexual development identified by Freud (see 1920, rep. 1975) in his psychoanalytic investigations.

A Type I power orientation corresponds to oral modes of expression, in which someone is concerned about being nurtured by another. In studying planners, one might find an example in someone who serves as a mayor's advisor on housing or economic development. While this person may consciously choose this role for the valid reasons that one can improve political decision making and promote the interests of the poor, unconsciously this person may like the role because it offers opportunities for being "fed" praise and compliments from a powerful person. This latter interest may make the advisor especially sensitive to the mayor's needs, although insecurity could lead the advisor to be so blindly loyal as to lose independent judgment.

Three points should be made about these assumptions. First, not every mayoral advisor has a Type I power orientation, and vice versa. Second, a correspondence between power orientation and psychological stages of development identifies one set of meanings for work but should not be interpreted reductionistically. For example, someone who serves as a loyal advisor to a powerful leader may enjoy gratifications analogous to those of a suckling infant, but the advisor's motives, opportunities, and impacts are considerably more complex. Third, unconscious motives for selecting professional roles in no way invalidate conscious motives, which may be quite respectable.

A Type II orientation corresponds to anal modes of expression, in which someone is concerned about regulating control over personally accumulated products. Among planners an example might be found in someone in data management. Data collection and analysis are essential for planning; however, some planners may be especially interested in collecting and organizing data as a way of satisfying personal needs for symbolic intellectual control over the planning process or turbulent social processes which the data represent. Any such special concern may, indeed, be particularly valuable in ensuring comprehensive data collection. However, if planners become fixed on controlling data as an inherent satisfaction, they may indiscriminately collect information just for the sake of collecting it, a potential waste of resources. Or they may resist releasing their data, reducing either the number of participants in planning or their knowledgeability.

A Type III power orientation corresponds to phallic modes of expression, in which someone is concerned about initiating assertions of strength toward others. Examples among planners might be found among community planners whose work includes political organizing. Consciously, these people may select these roles because they make it possible to confront powerful political or financial institutions. At the same time, these planners may unconsciously want opportunities to demonstrate their personal "potency" to others. This latter concern may contribute the quality of determination to their organizing activities, although insecurity could make practitioners repeatedly seek gratuitous opportunities to challenge others.

A Type IV power orientation corresponds to genital modes of expression, in which someone is concerned about joining with another or others in a relationship of mutual strength and identity. One might find an example of this orientation among planners who assist community groups in developing agreement on planning proposals or among those who attempt to negotiate resolution of land use disputes. Consciously, these people may select their roles for the reasons presented earlier— problem solving requires powerful collaboration. These people would differ from those of the Type III orientation in having an interest not simply in influencing others, but also in working with them. This interest might be shaped unconsciously by concerns about developing intimate work relations that either supplement or substitute for intimacy in the family. This latter concern may make planners particularly sensitive to others' expectations in collaboration; however, insecurity could lead practitioners to regard "comfortable" groups as ends in themselves, rather than as means to instrumental work.

McClelland's framework can be applied to problems in learning to act powerfully in bureaucratic organizations. The power orientation many planners and other bureaucrats find troublesome is the interpersonal Type III orientation. McClelland's conclusion that the orientations are developmentally linked would suggest that many of these people are concerned with mastering problems of the Type II orientation and that, until they feel confident that they have resolved Type II problems, they will be unable to begin to learn the Type III orientation.[16]

For example, planners may concentrate on information management more for unconscious reasons than for conscious reasons. They may be anxious about social changes and feel that collecting data about these conditions offers sufficient symbolic control over these processes to contain their anxiety. Or they may sense that they need to approach others and persuade or negotiate with them in order to make their information matter, but they may feel anxious about confronting others. In response, they may tacitly conclude that, if only they accumulate enough data, the data will "speak for themselves" and for the planners, so the planners can act powerfully without having to act personally.[17] The challenge to these people is to come to feel that they have both sufficient and appropriate control over their data. Then they and others may agree that they are collecting simply the amount and type of data reasonably required for problem analysis and are disseminating the data to those who need them. Until they develop such a stable feeling of Type II power, they may have difficulty thinking about, looking at, or attempting to learn Type III power.[18]

What is required for this transition is suggested by Erikson (1963), who characterizes the same four stages in psychosocial terms. Erikson describes the second stage as concerned with learning autonomy and the third stage as concerned with learning initiative. In order to master the second stage and to progress to the third, someone must learn to be autonomous without succumbing to feelings of shame and self-doubt. Erikson portrays the conflict of the second stage as "autonomy versus shame and doubt":

> This stage . . . becomes decisive for the ratio of love and hate, cooperation and willfulness, freedom of self-expression and its suppression. From a sense of self-control without loss of self-esteem comes a lasting sense of good will and pride; from a sense of loss of self-control and of foreign overcontrol comes a lasting propensity for doubt and shame. If, to some reader, the "negative" potentialities of our stages seem overstated throughout, we must remind him that this is not only the result of a preoccupation with clinical data. Adults, and seemingly mature and unneurotic ones, display a sensitivity concerning a possible shameful "loss of face" and fear of being attacked "from behind" [doubted] which is not only highly irrational and in contrast to the knowledge available to them, but can be of fateful import if related sentiments of influence . . . policies. (1963, p. 254)

In short, in order for someone to feel prepared to take initiative in acting powerfully toward others, this person must feel confident that he or she can act autonomously without encountering crippling shame or self-doubt.

Erikson's formulation of a conflict of "autonomy versus shame and doubt" helps explain the difficulties many planners and others experience in learning to act powerfully in a bureaucracy. Erikson intends the term "autonomy" to refer to a commonsense notion of personal independence. Sennett uses the term "autonomy" to refer to a specific type of self-sufficient authority, here postulated to be common to bureaucratic organizations. Erikson suggests that experiences of shame and doubt will interfere with an individual's acquiring autonomy and progressing to learn other capabilities, such as taking initiative. Here it is argued that autonomous authority generates experiences of shame and doubt for subordinates in such a way that they will not acquire autonomy in Erikson's sense and cannot challenge autonomy in Sennett's sense. In other words, acquiring secure feelings of autonomy (in Erikson's sense) is necessary to develop the initiative associated with a Type III power orientation, which is needed to act powerfully against the type of bureaucratic authority that Sennett defines as autonomy.

Shame

Shame, as Erikson indicates, involves a "loss of face." It involves exposure, particularly unexpected exposure, in which one is found "out of place." One is discovered to have done something incongruous or conspicuously inappropriate to a situation or a social role. What is upsetting about shame is discovering that some previously trusted people or personal assumptions were really unreliable. Unconsciously, the witness for shameful actions is identified with the ego ideal, the image of perfection toward which one strives. To varying degrees, adults transfer such images onto others in their everyday lives, including supervisors at work.[19] Thus shame is a loss of face before a parentlike figure whom one imagines to be a paragon of everything one wants to achieve. Further, this loss of face is experienced as a loss of parental affection and sustenance. Accordingly, anxiety associated with shame is not simply a reaction to shattered faith in specific social relationships; it is apprehension about the loss of continued social existence. Shame is not simply an emotional wound to some part of the psyche; people who are ashamed may feel totally mortified. Indeed, anthropological evidence from traditional societies shows examples of individuals literally dying after shaming experiences. Thus shame is a blow to one's entire identity; one may wonder whether life can go on after one is shamed (Lewis, 1974; Lynd, 1958; Piers, 1953, rep. 1971).

Autonomous authority controls by consistently generating experiences of shame for subordinates. Some of this shame would inevitably be evoked by any authority, for the ideology of Western industrial society holds individuals responsible for their relative status in society, and any subordinacy may be considered shameful. Regardless of what people have learned about structural obstacles to mobility in American society, most retain the feeling that, somehow, they are personally responsible for their success or failure. Because relationships of power and dependence are rationalized in images of authority and strength, being dependent is explained and experienced as being relatively weak, as having failed in individual striving for status. Hence in this society people tend to feel ashamed about being dependent on any one.

In addition, autonomous authority entails specific ways of engendering shame and using this shame to maintain control over others. Sennett describes this process:

> A person who has marshalled his or her resources, who is therefore self-controlled—this autonomous figure can discipline others through making them feel ashamed. Indifference to ordinary people has, of course, a shaming effect: it makes them feel they don't count. . . . shame has become stronger as violence has waned in Western societies as an everyday tool of discipline. . . . The shame an autonomous person can arouse in subordinates is an implicit control. Rather than the employer explicitly saying "You are dirt" or "Look how much better I am," all he needs to do is his job—exercise his skill or deploy his calm and indifference. His powers are fixed in his position, they are static attributes, qualities of what he is. It is not so much abrupt moments of humiliation as month after month of disregarding his employees [or subordinates], or not taking them seriously, which establishes his domination. The feelings he has about them, they about him, need never be stated. The grinding down of his employees' [or subordinates'] sense of self-worth is not part of his discourse with them; it is a silent erosion of their sense of self-worth which will wear them down. This, rather than open abuse, is how he bends them to his will. When shame is silent, implicit, it becomes a patent tool of bringing people to heel. (1980, pp. 92–93, 95)

In this way the bureaucratic experience of autonomous authority brings shame to subordinates. The following composite examples from planners' discussions of their work illustrate this aspect of subordinacy (Baum, 1983a). A staff planner confronting a self-sufficient, indifferent administrator is continually in the position of asking for something from someone who needs nothing. The subordinate must ask, first of all, simply for the administrator's attention. Planners quoted earlier emphasize the importance of this attention to their own assessments of their overall effectiveness. They want, above all, "responses from other people," "being taken seriously by my director," "indications that they are giving your views some considerations," just some "acceptance . . . from the powers that be." Any specific request for attention may be doubly shameful if the administrator implies in any way that the request is an interruption of more important administrative work, about which the subordinate can know and understand little.

Realistically, an administrator in a hierarchical organization is responsible for and knows about the work of many staff members, any of whom is unlikely to know much about the work of others, much less the coordination of efforts required to keep the organization running smoothly. In addition, when the division of labor within the organization is relatively specialized, it is particularly difficult for a planner in one area to picture or understand the organizational purpose which somehow incorporates his or her efforts along with others. The administrator may refer to organizational goals in such a general or abstract way that it is difficult to comprehend any concrete relationship between an individual's efforts and these goals. But the administrator's evident understanding of these goals presents one further sign of that person's strength and authority.

Beyond a plea for attention, any substantive request to an administrator induces shame because the subordinate is in the position of asking something from someone who cannot need anything in return. Discussions between an administrator and staff member normally focus on the subordinate, with the initial premise that the subordinate is needy.[20] Moreover, according to organizational rules, at various times a staff member may need the administrator's authority to obtain such things as money (in the form of an initial salary or raise, authorization for purchases of needed supplies, or reimbursement for work-related expenses), time (in the form of leave for vacation, conferences, or daily work breaks), status (in the form of initial rank or promotion, as well as designation to special positions), and space (in the form of size and location of work space).

For most workers these things are not simply abstract amenities, but are often central items in tacit negotiations of the psychological contract. Workers care not simply that they receive these things, but that they are strong enough to force their boss to give them. The asymmetry of autonomous authority thus inherently wounds their self-esteem. Further, it is hardly an exaggeration to say that the only other time in most people's lives when they had to depend on another for all these important things was childhood. Consequently, it should not be surprising if the situation unconsciously reminds bureaucrats of childhood experiences and leads them to react with feelings from those early situations. Thus subordinates may recall extreme dependency and feelings of helplessness. Insofar as they unconsciously feel these emotions once again at work, they may exaggerate any real differences in strength and feel additionally shamed about their still greater weakness.

Throughout any meetings between administrator and planner, the administrator is understood to have no needs that are legitimately discussable. The administrator's request for work is not an expression of need. Instead, it is simply an impersonal reminder that in order to be considered a responsible member of the organization, the subordinate must perform certain tasks. The administrator's assignment of responsibilities is not a request for assistance; it is a means of helping subordinates do their work so as to be acceptable to the organization.

The administrator's request to see the planner's work for evaluation is potentially the most shame-inducing situation of all. Even if organizational goals, roles, tasks, and techniques are ambiguous, in the end, the subordinate's work will be assessed for its fit to these ambiguities. Regardless of the administrator's actions, some incongruity—and some shame—are probable (Lipsky, 1980). In addition, contacts between the administrator and the planner may have been sufficiently infrequent as to increase the latter's uncertainty about the propriety and quality of the work. This lack of communication may contribute to differences in assumptions, so that whatever the planner presents will violate some of the supervisor's expectations. Also, this evaluation may be the supervisor's first opportunity to give a project concentrated attention, and thus present the first time the superior offers questions or suggestions. Execution of the assignment may have revealed some complexities not anticipated by the administrator, in which case, again, the administrator's expectations may not be met. On top of this, the administrator may want to demonstrate his or her expertise by identifying specific considerations that were not taken into account. Insofar as organizational work is highly specialized, the planner may be forced to accept the probability that he or she will never understand the place of the work in the organization and, therefore, will always make faulty assumptions.

Still, an evaluative meeting might present an opportunity for mutual learning, an opportunity, as one planner says, "to hassle down something." The discussion might be seen as evaluation of the work, rather than the worker. However, Argyris's (1982) research shows that, despite intentions to listen and advise, when supervisors see serious problems in work, they tend to criticize and blame subordinates in such a way that the subordinates react defensively. Hence a planner mentioned earlier draws a defiant conclusion about the evaluation of his work: "I am able to say it is a good piece of work, whether they like it or not." In the end, another asserts, "It goes back to my criteria, if it meets my criteria."

Many planners experience the evaluation as exposure in having performed inappropriately. They feel that they relied on faulty assumptions about their work and its place in the organization. Crucially, insofar as they have accepted an administrator's assignment as defining their responsibilities as competent practitioners, the administrator's disappointment shames them and undermines their personal identity. One planner suggests his personal hurt by saying that he does not look for administrative acceptance, "because if you look for it and don't find it, you just get grumpy."

These planners resemble the executives whom Maccoby (1976) studied. When he gave them Rorschach inkblots, their responses indicated that

> one of their most repressed feelings is humiliation at having to peform for others—from parents and teachers in childhood to the admired superiors at work—to be vulnerable and judged by them no matter how much the corporate policy emphasizes "respect for the individual." (1976, p. 117)

The corporate managers' responses to a questionnaire about personal difficulties sketch out a picture of the symptoms of shame anxiety. Many speak of fear, largely the fear of unexpected exposure for acting inapproprately before their superiors; 58 percent report often being restless, 48 percent acknowledge being anxious, and 28 percent say they have unwarranted fears. Most report some kind of detachment, which includes hesitance about appearing before others and doubt about their abilities: 61 percent say they keep their feelings to themselves, 59 percent say they have difficulty saying what they mean, and 35 percent say they avoid people. Many also report that they lack self-determination, that they are ashamed about asserting themselves, or that they doubt that they know what to assert themselves about. Fifty-nine percent say that they give in too easily to others, and 46 percent say they don't know what they really want (pp. 200–201).

Maccoby interprets these managers' fears in terms consistent with planners' statements. Above all, they worry that "Someone above them will decide they don't measure up and 'zap' them" (p. 201). Maccoby believes this fear has three sources:

> One type includes fears that something concrete will happen or not happen—that the project will not succeed, that a sale will not be made, or that one will be found out in a half-truth or lie. The second type is fear based on lack of knowledge. Many managers do not understand the business and their function in it. They are afraid that they aren't really needed and that someone will discover it. . . . The third type of fear is more pervasive, a constant gnawing anxiety not usually tied to any particular event or knowledge, but to loss of control. It is anxiety about losing one's cool, clutching at a crucial moment, looking bad, saying the wrong thing (and in some cases revealing repressed envious, cannibalistic, hostile impulses). (p. 202)

This constellation of feelings is reflected in the following comment by a public agency planner when asked what he looks for as an indicator of his professional effectiveness:

> No editing. When my foremost detractors don't question anything. There is not that much backslapping and congratulations. When I have done something decent, I get some congratulatory remarks from colleagues [but not necessarily from my director]. I don't need this all the time, but now and then it is nice. Often here you feel you are being taken for granted.

The final statement is perhaps the most important. This planner, as many workers in Chapter 3, complains that recognition is rare. As a consequence, it becomes quite important, perhaps more important than satisfaction of any internal standards. And yet the primary recognition he receives is negative. Accordingly, his criterion for success is avoiding criticism and the shame accompanying it. His resignation shortly after the interview suggests the mortifying quality of this shame.

If the planner does not leave the agency, the continuing shameful experience of subordinacy may discourage any interest in acting powerfully toward either the administrator or others. As a result of continual experiences of shame, a staff member is likely to feel increasingly anxious about performing and concentrate on efforts intended to allay evaluative anxiety. The staff member will pay more attention to cues about the administrator's thoughts and preferences, increasing dependency. Not only is the subordinate shamed from seeking autonomy, but the thought moves further and further from consciousness. These are hardly conditions for a satisfying psychological contract.

Doubt

Feelings of shame may be accompanied by closely related feelings of doubt. Doubt is experienced as a loss of confidence in one's assumptions about the world and, centrally, a loss of confidence in oneself. Usually, when one has experienced shame about something, one comes to doubt one's ability to act in any way not shameful. Conversely, when one experiences doubt about one's assumptions or abilities, then one usually becomes ashamed of this incapacity.

The area most susceptible to doubt when confronted by autonomous authority in a bureaucracy is work competence. The importance of the psychological contract with the organization indicates that the significance of work extends beyond the material rewards it provides. Culturally, work activities are highly valued. These activities occupy a major portion of most people's lives. Work provides one of the few opportunities for people to be personally creative. People tend to choose occupations that are consistent with their perceived abilities and personality orientations, and they expect a job to provide them with an opportunity both to use their abilities and to express their personality (Holland, 1973). Thus the ability to satisfy personal standards in performing competently at work is intimately tied to self-esteem.

Autonomous authority tends to act against the need for self-esteem by challenging a subordinate's judgment about his or her competence. This process of generating self-doubt is illustrated by the following composite examples from planners' observations (Baum, 1983a). To begin with, ambiguities in problems, organizational policies, and evaluative criteria almost inevitably plant seeds of doubt in subordinates' minds regarding their ability to satisfy superiors (Thompson, 1961). In this situation, the autonomous administrator's strength is defined as an expertise in the subordinate's work domain. Specifically, it is an expertise about work acceptable within the organization. As the planner continually seeks a sign of the apparently indifferent administrator's preferences, the planner may unconsciously "fill in" the supervisor's expectations by identifying the supervisor with the planner's own ideal accomplishments, or ego ideal. As a result, the superior may seem to have extremely high expectations, with which the subordinate intimately identifies. Satisfying oneself and satisfying the autonomous authority become virtually the same thing.

Insofar as the administrator is the planner's primary contact with the rest of the organization, it is difficult for the subordinate to develop an independent realistic view of appropriate work. The infrequency and ambiguity of contacts with the administrator create a vague, confusing image of the organization. The planner may receive an assignment from the administrator, may carry out the assignment in a well-considered manner, may submit the finished product to the administrator, and yet have little conception of the rest of the organization that the administrator represents. It may be unclear what events generated the assignment, and it may be unclear what happens to the finished product. This mystery about the origin and final destination of the work further perplex a planner about how to assess his or her work.

A fusion of the subordinate's identity and the administrator's preferences produces in the planner a strong sense of doubt about both work competence and the ability to assess this competence. A local planner provides an example of planners' repeated doubts about the assessment of their efforts:

> In terms of whether or not the work I do is [technically] *right*, I don't have all that many qualms about it. Whether or not it is *useful* [in the organization] is where I am on softer grounds where assessment is concerned.

Bureaucratic isolation fans this doubt, as another observes:

> Sometimes I don't have a measure of comparison. In new areas I am so isolated, I don't know what others are doing. There is not a rich environment. In school you are usually exposed to a great variety of other people. Here your work is done, but you don't know [whether those who read it are competent to judge it].

For some, the solution to this doubt is simple: the criterion for effectiveness, as one worker says, becomes, "in general, the acceptance which I get from the powers that be." These responses by planners resemble those found by Kahn et al. (1964) in studying stress among a national sample of bureaucratic workers. Their study found that pervasive ambiguity in bureaucracies contributes to self-doubt.

This feeling disables planners and other bureaucrats in three ways. First, they tend to doubt that they have the right to make requests of supervisors with autonomous authority, either for specific objects or simply for attention. Second, staff members tend to doubt that they have the ability to make any requests effectively. Third, whenever staff members do become involved in discussion with administrators, they tend to doubt the correctness of their own position. Because discussion focuses on the subordinates' work, with all these doubts, there is never consideration of any weaknesses of the administrator, and the autonomous authority remains intact. This doubt gives rise to shame and creates additional anxiety for the staff member. It stimulates even greater attention to the administrator for hints about appropriate standards for competent work. Not only does the subordinate member doubt that he or she should seek autonomy, but the subordinate doubts that he or she could exercise autonomy.

THE DOUBTFUL DEFENSE

Yet planners' doubts do not stop here. Many express doubts about whether there is a defining core of planning expertise or a guiding body of ethics for their profession, whether other planners are a relevant reference group, whether as a group they are sufficiently distinctive to be licensed, and whether they should act through a professional association (Baum, 1983a). Some of these doubts are expressed by a public agency planner:

> I don't know what planning is. It is hard to say what the expectations are of something you cannot define. I think my outlook is guided, rather, by some concrete expectation, by my training, and my reading of the planning literature. I have internalized planning attitudes. I don't know if they are listed anywhere. I feel that my attitudes in many ways are attitudes that planners have, but I am not sure I could write them down. Maybe the American Institute of Planners' codes . . . Sometimes I go back to the A.I.P. Code of Ethics. . . . I use the A.I.P. Code of Ethics as a crutch, to affirm my position. I am always falling back on my own moral outlook.

Bewilderment about whether there is a community of interest among planners is expressed by another practitioner:

There is a reference group, but I am not sure that I can indicate what it is. It is not the planning profession, and it is not the A.I.P. It is to some degree the people with whom I work.

These doubts lead to doubts about whether planners should be licensed, as a county planner indicates:

I think it is preposterous. I don't know how the hell you can license somebody when you do not have criteria for what it is or what they do. . . . I don't think planning is a clearly defined enough discipline to draw a test to say that one guy is a planner and another is not. . . . Most of the effective procedures which I see are adopted from management. But there is no body of knowledge, as there is in the field of economics or engineering, and, therefore, you can have people like the planning director I worked for in Pennsylvania, who is one of the most skillful planning directors there is, and he has only a bachelor's degree in political science. That is probably why he is a good planner. He did not muck up his head by going to planning school.

Doubt and Shame

So much doubt would clearly disable practitioners from beginning to develop a Type III power orientation. Questioning the meaning of one's work would certainly discourage assertions of authority over others. Still, such extensive doubt is surprising among practitioners who say they would like to have more power. The pervasiveness of doubt suggests either that the anxiety associated with asserting Type III power is intense or there may be hidden rewards from doubting, or probably, both.

Clinical evidence about doubt supports this speculation. Freud (1930, rep. 1965) notes that when people doubt whether they actually dreamed parts of dreams they remember, usually they are attempting to avoid conscious recognition of unconscious thoughts which would make them anxious. Sullivan (1956) observes that doubt is often a means of resisting a certainty that would induce anxiety and thus defends against the anxiety. In this way, doubt about the existence of well-defined planning expertise and ethics might protect a planner from the anxiety associated with acting competently and ethically to assert his or her position in conflict with others.[21]

Doubt is closely associated with anxiety about being shamed by persons in authority who are assumed to be stronger intellectually and in other, poorly defined ways. Planners are ashamed they cannot assert themselves either professionally or politically in relation to these bureaucratic superiors. Thus if a subordinate contemplates challenging someone in authority, an action that might be dictated by conscious intentions to exercise Type III power, this thought is unconsciously associated with the humiliating anticipation of certain defeat—really, psychic annihilation—by the stronger authority. Significantly, doubt may defend a subordinate against defeat and increased shame by deterring the challenge of a superior. It may prevent a worker from even considering challenging the superior. Although a worker may feel somewhat ashamed about the failure to assert his or her interests, the worker may feel still more relieved by the security of a nonconfrontive position. In this respect, doubt is a successful unconscious strategy to avoid the Type III power orientation and the anxiety it arouses.

In addition, doubt about the possibility of challenging a superior on important issues presents a means of taking revenge on the organization that constructs such shameful predicaments, by completing an unconscious syllogism. Planning requires rational confrontation of alternative views; bureaucracy does not permit such confrontation; therefore, bureaucratic organization and authority have no relevance to planning. They are irrational; they are "political"; they are immaterial. Like counterambiguity, persistent doubt is an effort to annihilate and punish an organization that apparently denies a satisfying psychological contract.[22] Perversely, extended assertions of inferiority, doubt, and shame may unconsciously represent the one way in which a subordinate can control a situation otherwise invidiously out of control. Further, shame may help to preserve a sense of separate identity—failure is humiliating, but it is personal.[23]

Thus a subordinate who reacts shamefully toward autonomous bureaucratic authority may fail at conscious intentions to exercise Type III power. At the same time, the subordinate may unconsciously capitalize on shame and doubt to devise a defensive strategy toward avoiding any assertion toward others. When a subordinate subsequently thinks about acting powerfully to influence decisions, the worker is likely to beat an unconscious retreat, or regress, from thoughts of Type III power to seemingly safer power orientations that have been successful in the past. A planner, for example, may turn back toward a Type II orientation and focus on collecting data, as many planners do. Or a planner may move back toward a Type I orientation and focus on loyally advising a superior, avoiding, as an earlier bureaucrat emphasized, any appearance of climbing on the superior's back. Although these orientations may be fruitful for solving problems, if planners unconsciously choose these positions out of anxiety about the Type III orientation, they may make ineffective choices.

Shame and Guilt

How are pervasive feelings of shame and doubt related to earlier evidence of subordinates' guilt about acting assertively toward persons in authority? Subordinates perceive and respond to two images of bureaucratic authority. The image in Chapter 2 portrays the authority as a moral paragon. The image here depicts it as superiority in competence or strength. Confronting each image arouses a different type of anxiety. Challenging the moral paragon contributes to guilt anxiety. Challenging—or often more accurately, disappointing—the superior strength contributes to shame anxiety. Each type of confrontation and its imagined consequences is associated with specific unconsciously remembered experiences and admonitions from earlier life. The qualities and power of a moral authority reflect a subordinate's unconscious assumptions about the superego, or conscience, constructed from childhood. The qualities and power of a strong authority reflect assumptions about the infantile ego ideal, or ideal aspirations.[24]

The two types of anxiety are dynamically related (Lewis, 1974; Piers, 1953, rep. 1971). Someone may unconsciously attempt to avoid a strong feeling of guilt for a moral offense by reinterpreting the real or imagined transgression against authority as a shame-inducing offense. For example, if the perception of bureaucratic authority in moral terms contributes to guilt for intentions to act powerfully, a subordinate may tacitly choose to emphasize the superior's strength and performance expectations, the disappointment of which may bring shame. Thus a subordinate might unconsciously reason in the following way, without being aware of thinking any of these things:

> If I attempt to challenge my boss' weak ideas on this proposal, I might prevail, and the proposal would be improved. However, I would feel guilty for challenging my boss; it would be like taking on my father, and defeating him. I could imagine his beating me severely for even considering that. I really want to avoid these thoughts. Instead, I will think of myself as weak, with the result that there is no risk that I could challenge my boss, hurt him, or be punished by anyone. Once I think of myself in this way, I have difficulty thinking of many good ideas regarding the proposal, the proposal won't be very good, and I feel ashamed that I don't have the kind of good ideas which my boss seems to have, but feeling ashamed is better than all the trouble of worrying about confronting, feeling guilty, and being punished.

Although shame is painful, the subordinate may consider it a momentary relief from the alternative of guilt. The passivity of shameful actions may seem preferable to the risks of more aggressive guilty actions. And even though shame is a totally mortifying experience, it depends on the real or imagined continuing presence of an authority who witnesses one's weakness. In contrast, guilt adheres only to a single action, not the whole person, but it pertains to the transgression of an abstract moral tenet. The company of the shaming witness may seem more secure than the solitude of guilt.

Menzies' (1975) description of nurses' "reduction of the impact of responsibility by delegation to superiors" portrays a similar process. In the hospital she studied, junior nurses who are anxious about making mistakes with patients and harming them unconsciously redefine the authority of the supervisors. Formally, these senior nurses may be considered strict evaluators who would punish subordinates for therapeutic errors. However, junior nurses tacitly reconstrue their superiors as highly competent practitioners and think of themselves as novices who know rather little. Significantly, the senior nurses tacitly agree to this reinterpretation. Thus with this transformation, the junior nurses choose the risks of being shamed for ignorance as preferable to the risks of guilt for taking initiatives in patient cases.[25]

Alternatively, shame may turn into guilt. For example, when someone unconsciously chooses inactivity and inferiority feelings in order to avoid guilt, this person may feel ashamed of the passivity and turn to new, exaggeratedly aggressive actions in order to reinstate dignity. These actions may give rise to new guilt anxiety. More generally, any feelings of shame may evoke rage against the person who shames by comparison. At the same time, tacit recognition of the irrationality of hostility toward someone who offends simply by being superior will probably lead to suppression of the anger. The shamed person is likely to feel guilty for unjustly wanting to harm the other, particularly someone who is supposed to nurture. Thus a subordinate might unconsciously think about these feelings in the following way:

> I don't like feeling that I never have as much to contribute as my boss does; I feel ashamed of myself. In fact, I could offer some really good ideas if I just sat down and concentrated on it. I might even begin with the weaknesses in his ideas and develop specific improvements on them. In general, I really don't like the way he always seems superior; it makes me angry. I'd like to knock him off his pedestal once or twice. Still, it's really irrational to think like this—he's never really hurt me, and it's his job to have

> good ideas. How could I think so much about getting him? I'm really being unreasonable, and I should watch myself, so I don't offend him and get into trouble. I feel guilty that I even have such thoughts.

In this way, guilt about rage toward superiors may supplant the initial shame.[26] At least for the moment, guilt is preferable. Perhaps the partialness of guilty offenses is preferable to the totality of shameful ones. In addition, as the next section suggests, guilty offenses are easier to remedy than shameful offenses.

At different times, bureaucratic workers may experience both guilt and shame in relation to their superiors. Which experience dominates at any time depends in part on the characteristics of supervisors and of subordinates. For example, a supervisor may moralistically evaluate subordinates or may emphasize their moral obligation to adhere to agency norms. Alternatively, a supervisor may attempt to teach subordinates or may seek to show up their performances. In addition, the balance of shame and guilt also reflects a worker's early experiences and ensuing personality orientation. Some people are more sensitive to moral issues or transgressions, while others respond more quickly to possibly shaming situations (Lewis, 1974; Lynd, 1958).

What is the contribution of the bureaucratic psychological structure to the balance of shame and guilt? Bureaucratic authority, in the context of American beliefs about personal achievement, work, and success, appears to be most commonly interpreted by workers as a potentially shaming relationship. Work is not primarily an issue of obedience to a superior moral authority. It is above all an expression of personal competence, and workers resent others' controls over such self-expression and do not believe their hierarchical superiors know much, if anything, more about their work than they do themselves. Although anger against superiors and guilt for it have many causes, much of the basis for these feelings appears to be workers' self-defense against the painfully shameful experience of dependency on autonomous bureaucratic authority. Guilt, after all, may be preferable to shame, because only someone who is strong and capable could commit an offense warranting guilt.

SHAME AND GUILT AS INSTRUMENTS OF SOCIAL CONTROL

Bureaucrats may unconsciously choose guilt as a defense against shame because of the more attractive self-image offered by moral transgressions. In addition, workers may prefer definitions emphasizing moral expectations and guilt because such offenses are more remediable. Piers summarizes the differences in transgressions:

> Whereas guilt is generated whenever a boundary (set by the superego) is touched or transgressed, shame occurs when a goal (presented by the ego ideal) is not being reached. It thus indicates a real "shortcoming." Guilt anxiety accompanies transgression; shame, failure. (1953; rep. 1971, p. 24)

Importantly, transgressions inducing guilt are specific actions more or less deliberately taken by someone, who may be, except for the specific transgression, morally intact and acceptable. In contrast, unanticipated experiences of shame come in response to a person's being exposed as unacceptably "out of place"; the entire person is mortified. As a consequence, guilty actions are remediable, whereas shameful actions are not. Remediation of guilt entails atonement for a specific offense. In contrast, shame is difficult to remedy because a whole person has stepped out of place and can never undo that condition (Lynd, 1958).

Thus situations that are defined as measures of performance or strength are more likely to control subordinates than situations defined as measures of moral propriety. Offenses against the former are virtually irremediable, and the attendant anxiety may be greater. This suggests that bureaucratic administrators who are concerned about maintaining control over subordinates are likely to encourage definitions of subordinacy in terms of competence rather than moral loyalty. This approach is consistent with the Weberian ideal-typical rationale for bureaucracy. Argyris and Schön (1978) offer evidence that such a definition of bureaucratic situations is attractive to administrators.

In response, subordinates may hold two alternative definitions of the hierarchical relationship. If subordinacy is apparently inalterable, subordinates may seek to define situations of their subordinacy in terms of moral loyalty rather than competence. In that context, offenses against authority may be treated as remediable rather than irreparable.[27] Alternatively, particularly if the subordinates have some power, they may attempt to redefine the role of the authority. Gouldner (1964) observes that subordinates may attempt to control superiors by instituting new formal standards to which the superiors are expected to adhere. Such a maneuver replaces an authority of top management based simply on organizational position with an authority tied to expertise, an ability to perform according to specific standards. This move is an effort to subject hierarchical superiors also to potentially shameful experiences. Both approaches are efforts to redefine bureaucratic situations in which subordinates alone are likely to be measured in shameful ways.

CONSEQUENCES FOR PROBLEM SOLVING

Even if bureaucratic workers appreciate the strategic importance of powerful action in solving problems, the structure of bureaucratic authority may frustrate powerful intentions. The autonomy of bureaucratic authority tends to engender shame and self-doubt among subordinates. These feelings both discourage attempts to act powerfully and hinder thinking about reasons for powerful action. They increasingly divert staff members from doing their work to wondering about their administrators' expectations. In an attempt to derive some satisfaction from work, a staff member may turn from bringing about substantive results outside the organization to getting compliments or praise (recognition) from others within the organization—ideally from the chief administrator, but if not, from anyone available. Workers who doubt their own efficacy and feel humiliated by organizational authority may not want to think about how power is exercised or what they might do to acquire it.

In response to their shame, subordinates may become hostile toward others, perhaps beginning with supervisors and moving to colleagues. The resulting guilt and fear of retribution will limit their efforts to act powerfully toward others. Any hostile assertions are likely to be self-defeating and are unlikely to elicit collaboration.

This analysis has focused on the problematic transition from a Type II power orientation to a Type III orientation, in which workers may be at ease challenging others. However, the powerful action required for collaborative problem solving requires an expression of a Type IV orientation. Collaboration entails an investment in defending the identity and interests of a long-standing or temporary group.[28] Workers must feel that their identity is secure from individual challenge before they can invest in allied activity with others. Examples of such collaboration include organizing or working with citizen groups, as well as developing or working in coalitions or networks with people in other agencies or interest groups. As McClelland observes, the power orientations are linked in a developmental logic. Difficulties in learning a Type III orientation are also obstacles to a Type IV orientation. Thus the bureaucratic psychological structure which fixates workers on overcoming shame and doubt inhibits not only autonomy, but collaboration as well.[29]

There are planners and other bureaucratic workers who act powerfully and collaboratively. They lead and participate in complex problem-solving efforts. The analysis here does not deny the existence of these practitioners but does suggest the possible psychic difficulty and costs of success for some of them. In order to act powerfully, they must invest considerable energy in a major psychological enterprise not evidently directly related to their work. They must learn to differentiate between the manifest and the unconscious realms of their organizations. In particular, they must learn to distinguish between rational arguments against using power and the urgings of anxiety associated with guilt or shame in relation to imaginary authority figures. In this way they may successfully unravel the double bind that both encourages and forbids the exercise of power.

NOTES

1. It should be clear that the issues, and their confusion, are similar for administrators as well. The aim of administrative practice is commonly formulated as the infusion of disinterested rationality into decision making and the restriction of self-interested political considerations. For an excellent discussion, see Denhardt (1984).

2. Lasch (1984) has painted the cultural context for this view of power. In writing of "the normalization of crisis," he describes the ways in which many have taken the potential for nuclear holocaust to mean that any political action is futile and any assertion potentially catastrophic.

3. For an analysis of double binds, see Bateson (1972) and Laing (1972). For examples drawn specifically from bureaucratic organizations, see Argyris and Schön (1978).

4. Chasseguet-Smirgel (1985) offers an insightful psychoanalytic view of infants' reactions to the discovery of their parents' superior strength and sexual competence. Her book is the best description of the development of the ego ideal. In her account she emphasizes that the extraordinary qualities attributed to parents originate in infants' imagined recollections of their own omnipotence in the womb.

5. One infantile way of viewing the world, which Klein (1959) labels the "paranoid position," involves splitting the world into good and bad objects, or qualities, such that the infant regards itself as all good and the parent as all bad. The expulsion of bad qualities brings some satisfactions, even though it creates external threats. Memories of these infantile satisfactions lead adults to deal with threateningly ambiguous situations by denying any bad qualities in themselves and attributing all bad qualities to others, including leaders. Such a

defense process would lead to the perception of leaders as strong but bad, but this unconscious thinking is different from the conscious judgment to which Sennett refers.

6. Although Sennett (1980) has formulated the concept of autonomous authority, he has not identified it exclusively with bureaucratic authority.

7. For a description of introjection, see Schafer (1968). For an analysis of introjection in groups, see Freud (1921, rep. 1959). Freud (1946) gives particular attention to introjection as part of "identification with the aggressor." Foucault (1979) describes introjection of authority among inmates in Bentham's "panoptic" prison. Thompson (1961) offers observations about transference and introjection in bureaucratic organizations.

8. Some work is statutorily initiated. For example, a staff member working with a zoning board may be automatically required to prepare recommendations on zoning appeals. Some responsibilities are "assigned" by community groups. For example, a community planner with a public agency may be expected to help neighborhood associations develop proposals for whatever problems they identify. Still, in both cases the staff member is assigned to the position by an administrator and is subject to whatever directions the administrator gives about bounds to discretion, proper proposals or recommendations, and the like.

9. Significantly, almost all interviewed subordinate staff members in public agencies appear to reject implementation or constituency satisfaction as indicators of their effectiveness and, instead, focus on compliments or praise from agency colleagues as the primary indicator of their effectiveness (Baum, 1983a).

10. Mechanic (1962) argues that administrators' dependency on subordinates for implementation of organizational policy—that is, for getting work done—gives subordinates some independence. Lower-level staff have unique information about technology and clients, as well as access to clients and often, peers in other organizations. Consequently, subordinates know how to subvert administrative regulations and how to innovate. At least tacitly, Mechanic suggests, managers accede to these resources. The important questions Mechanic raises are how much discretion supervisors are willing to give subordinates, and under what conditions, with how much information, supervisors will evaluate or discipline subordinates.

11. See Thompson (1961) for a discussion of such "bureautic" behavior.

12. Gouldner (1964) offers just this explanation for administrators' creating or endorsing autonomous roles.

13. Significantly, Milgram (1974) found that variations in the personality of authority figures did not affect the responses of subordinates in his experiments. He concludes that the situation of authority is more influential than the personality of the authority holder on the response of subordinates.

14. See Hodgson, Levinson, and Zaleznik (1965) for numerous examples of this phenomenon. The self-fulfilling quality of transference is familiar to clinical workers. For example, see Kernberg (1965), Orr (1954), Racker (1968), Reich (1951), and Winnicott (1958).

15. The interpersonal power interest in Type III contrasts also with the interpersonal interest of Type I, in that the latter is concerned with another's control over oneself. However, consideration of the Type I orientation is not central to the discussion that follows.

16. An important empirical question corresponding to the hypothetical examples of planners with different power orientations is under what conditions individuals are capable of advancement to higher stages. Psychoanalysts speak of "character" to describe patterns of habitual responses to events. Thus the four hypothetical examples might represent, respectively, persons with an oral character, an anal character, a phallic character, and a genital character. What this would mean is that each example might represent someone who, even though an adult, is for some reason fixed, or stuck, at an earlier stage of development and way of reacting to the world; the person might move back to prior stages but could not advance. (The exception would be the fourth character type, which represents ideal mature development.) In these cases, little variation in the social environment, such as entry into or departure from a bureaucracy, would influence their normal mode of response to others.

Probably character traits do limit many bureaucrats' repertoires of organizational actions. Still, the empirical question here is whether how someone commonly acts in a bureaucratic organization is the most advanced way that person is capable of in other circumstances. This book suggests that, even though people may have characteristic modes of action, bureaucratic psychological structures may lead people to regress, or move back, to orientations beneath their capacity, or act less competently in bureaucracy than outside. Accordingly, modifications in organizational social structures might enable members to advance in power orientation, at least up to the level of their primary character traits. See Fenichel (1945) for a discussion of character, regression, and development.

17. In this latter case, emphasis on a Type II orientation could be regarded as a "regression" from Type III interpersonal contacts, or a psychological retreat from a later developmental stage that is unsuccessfully mastered into a more successful earlier stage. This retreat would defend the practitioner from the anxiety of the interpersonal contacts. For a discussion of the defense mechanism of regression, see Freud (1946) and Freud (1936; rep. 1963).

18. The question might be raised, how, if the power orientations represent developmental stages, a planner, as suggested earlier, might engage in Type II and Type IV orientations without engaging (being able to engage) in a Type III orientation. There is no definitive general answer for all planners, but an explanation that may refer to some planners is suggested by Erikson's (1963) discussion of the development of identity in adolescence, a period corresponding to the initiation of a Type IV orientation to power. During adolescence, individuals seek to develop coherent identities in terms of social relationships and worldviews. For many adolescents, Erikson observes,

> the adolescent mind becomes a more explicitly ideological one, by which we mean one searching for some inspiring unification of tradition or anticipated techniques, ideas, and ideals. And, indeed, it is the ideological potential of a society which speaks most clearly to the adolescent who is so eager to be affirmed by peers, to be confirmed by teachers, and to be inspired by worthwhile "ways of life." (1963, p. 130)

In initial experiments with identity, early adolescents may invoke particularly strong ideological principles or ideals as a means of setting themselves off, intellectually and socially, from others. As anxiety about differences diminishes, the adolescent's invocations of ideologies may soften or develop more social fit.

With regard to planners who may move between Type II and Type IV power orientations, one might hypothesize two patterns. One group is, in fact, capable of a Type III power orientation but chooses to emphasize nonpersonal expressions of power, perhaps in response to professional norms against overt political action. Another group, however, may in some ways resemble early adolescents in terms of their use of ideology. They may make recommendations on the basis of principles of the public interest or rationality that are rarefied and highly abstract and do not reflect appreciation of the complexity of the social world to which the principles might refer. Examples of the invocation of such notions of the public interest are reported by Howe (1983). Thus these planners may give the appearance of Type IV power orientation without really having mastered either the Type IV or Type III power orientation.

19. Kohut (1971) calls this transference an "idealizing transference" and describes its manifestations in clinical settings.

20. In reality, as Mechanic (1962) notes, an administrator and a subordinate are usually interdependent, needing assistance from one another. For example, an administrator is helpless without the work of his or her staff. However, this summary of many planners' observations emphasizes how this relationship is experienced from below. This asymmetric perception may be encouraged by the managerial style reported by Argyris and Schön (1974, 1978).

21. It should be clear that there are valid intellectual issues in planners' uncertainties about the professional status of planning. See Baum (1983a) for an examination of those issues. In this respect, planners share the uncertain professional status of many occupational groups which started out in bureaucratic roles. Planners, administrators, policy analysts, management scientists, and others began work as organizationally defined subordinates, rather than as relatively independent practitioners like physicians or lawyers, with considerable discretion in relation to clients. See Larson (1977) for an insightful analysis of the situation of organizationally employed engineers. The main implication of the analysis in this book, however, is that not all of practitioners' doubts reflect simply intellectual issues.

22. Fenichel (1945) argues that obsessive, or persistent, doubts about a situation may express doubt about the efficacy of other efforts to undo the situation, and thus provide renewed support for those efforts.

23. See Piers (1953; rep. 1971) for a discussion of this masochistic experience of shame.

24. This distinction is drawn and elaborated by Piers (1953; rep. 1971).

25. Piers (1953; rep. 1971) offers the following description of the general process of substituting shame for guilt:

It is an almost universal occurrence in our culture that impulses (or acts) of aggression generate a sense of guilt which results in their inhibition. This inhibition frequently spreads from destructiveness proper to assertiveness and in more pathological cases, to activity as such. The resulting passivity brings about a conflict accompanied by the anxiety signal of shame. (p. 32)

26. Lewis (1974) describes the phenomenology of shame-rage in the following way:

When humiliated fury is evoked, the self becomes the target of hostility. Expressing humiliated fury toward the apparent source of the humiliation is blocked by guilt for unjust anger; it is also blocked by positive feelings for the source of the humiliation, which co-exist along with the hostility. (p. 87)

27. For a powerful literary example of efforts to convert a shameful offense into a guilty one, see Kafka's (1930; rep. 1974) portrayal of Amalia's secret in *The Castle*.

28. Making recommendations on behalf of the public interest may express a Type IV orientation and may contribute to the solution of problems. However, the social activity of collaboration is essential for identifying potential solutions and developing commitment to them.

29. In Erikson's framework, collaboration requires a secure identity and success in developing mutuality. For a provocative formulation of stages of organizational development corresponding to Erikson's stages of personal development, see Torbert (1974/75).

5

Escape and Compensation: Dramatic Examples

The play's the thing
Wherein I'll catch the conscience of the king. (William Shakespeare, *Hamlet*)

We're all in this alone. (Lily Tomlin)

Man can will nothing unless he has first understood that he must count on no one but himself; that he is alone, abandoned on earth in the midst of his infinite responsibilities, without help, with no other aim than the one he sets himself, with no other destiny than the one he forges for himself on this earth. (Jean-Paul Sartre, *Being and Nothingness*)

When responsibility and authority are unequal and ambiguous, many workers feel ambivalently drawn both to seek recognition for their competence and to avoid recognition for fear of being blamed for problems they cannot control. As a defensive half-measure, they may attempt to confuse others about their responsibility by creating added ambiguity, but they lose the possibility of being recognized for their positive accomplishments. When at the same time workers regard organizational authority as autonomous, many feel deeply inadequate and ashamed about their weakness. They feel unable to assert themselves to their supervisors—for example, to insist on recognition and reward—and they increasingly doubt that they can do work worthy of recognition. Thus these predicaments make many bureaucrats ambivalent about wanting responsibility and confused about their ability to act authoritatively. Nevertheless, there is organizational work to be done, and they are asked to do it.

In response, bureaucrats may fashion a psychic "double identity." Overtly, they work on organizational problems, collecting information and meeting with people as required. Covertly, they imagine themsleves escaping and engage in activities that symbolically remove them from the organization's stresses. In fantasy they create an alternative organization that comes closer to providing a satisfying psychological contract. A county planner offers an example in assessing his work:

> I don't feel strong at all. My interest is urban design, but I haven't done too much urban design because I went up in my profession. Because I work with local governments, I was pushed toward administrative aspects of planning, which I don't like. [It is never hard for me to assess my work,] but I never tell anyone. I am a dreamer, but at the same time I am very realistic. . . . I wish I had my own consulting firm.

The fantasy of the urban design consulting firm balances the reality of the agency where he is a middle manager.

Some bureaucrats elaborate their wishes to escape into more full-fledged "dramas." These dreams may be daydreams, plans for the future, or at times actual "plays" that are performed alongside formal organizational activities. These dramas serve two purposes: to provide sufficient satisfactions to make it worthwhile for people to work in stressful predicaments; and if this fails, to provide an alternative realm into which to escape.

Recent organizational observers have given new attention to the dramatic aspects of bureaucratic life (for example, Mangham & Overington, 1983; see also Burke, 1969; Goffman, 1959; Thompson, 1961). Analysts of "organizational culture" point out that many organizations have mythlike stories that represent corporate history, symbolizing the company's purpose (see Deal & Kennedy, 1982; Martin, 1982; Martin, Feldman, Hatch, & Sitkin, 1983; Pettigrew, 1979; Pondy, Frost, Morgan, & Dandridge, 1983; Schneider & Shrivastava, 1984; Smircich, 1983). These accounts emphasize that the meaning of organizational actions is always a matter of interpretation, that organizational reality is "socially constructed" (Berger & Luckmann, 1967). Further, in most organizations certain dramatic portrayals of the organization come to dominate members' tacit assumptions or at least are recognized as orthodoxy.

The dramas examined in this chapter resemble stories in organizational culture in formulating organizational purpose but differ in three key respects. First, these dramas are individual constructions rather than learned organizational beliefs; second, they focus on the worker rather than the organization; and third, they represent responses to the failure of organizational culture, its ultimate incredibility. Members create alternative images of the organization, or images of alternative organizations, when corporate orthodoxy does not satisfy their needs to feel competent.

The ways in which workers create their dramas may be understood by comparison with children's play. Although more abstractly, adults "write," "direct," and "produce" dramas the way children play with toys. Children create toys, or designate objects as toys, in order to work through tensions between themselves and their environment.[1] Winnicott (1971) designates the first prototypical toy a "transitional object." Perhaps a blanket, it is the child's first object not wholly part of him- or herself, and yet it is not fully separate either. In this sense it is transitional. The infant's purpose in experimenting with this toy is to make sense of, and gain some control over, the mother's separation from him- or herself. By manipulating the toy often enough, repeatedly holding it and putting it aside, the child can gradually become more comfortable about the disturbing physical separation of the mother. Although the child may only reluctantly accept the mother's physical absence, the child learns that the mother will return, that she is usually somewhere nearby, and that thoughts of her may substitute for her actual presence.[2]

Workers' dramas, while related to more complex phenomena than children's play, share many of the same characteristics and purposes.[3] First, they are concerned with problematic, anxiety-evoking relations between the individual and the environment. Second, the dramas aim to fulfill wishes for security. Third, the dramas are transitional phenomena. In an important sense, their "authors" regard them as neither fully part of them nor fully separate. Their roles as "producers" of the dramas and "actors" in them may be blurred. In this way, the drama affords a way of experimenting risklessly with alternative resolutions of traumatic work situations. Fourth, closely related to the third characteristic, dramatic production is independent of the control not only of external reality, but also of the ego ideal and the superego. This means that the "authors" do not have to worry if their dramas fail to satisfy great ambitions or violate moral prohibitions. Dramatic activity may be free of shame, guilt, and related anxiety. Fifth, none of these activities would be possible if much of the creative work did not take place unconsciously.[4] When dramatic efforts succeed, they offer workers security without anxiety, and they may find dramatic realms more attractive, if not more "real," than formal organizational activities.

This chapter examines two workers' dramas. Each is concerned with a problematic aspect of the bureaucratic environment. The first focuses on responsibility, the second on authority. In their thinking, these people associate responsibility and authority with basic ontological categories: meaning and causality, respectively. Responsibility is concerned with various meanings of one's position and of bureaucratic relations, such as what workers expect of one another and how they respond to one another. Authority is concerned with causality: if one takes particular actions, how will others respond, and can one accomplish one's aims?

Strictly speaking, such a constrast between meaning and causality is artificial. In the social—as contrasted with the biological—world, causality is impossible to establish. Rather, people act on the basis of perceptions, of the meanings of their actions, and others respond in ways meaningful to them. Thus both meaning and causality are matters of perception.[5] Nevertheless, in everyday thinking, bureaucrats, as others, distinguish between the two.

These workers' dramas depict bureaucratic worlds in which meaning and causality, responsibility and authority, are problematic. As playwrights, the workers seek to establish new, reliable meanings and dependable causal relationships.

THE ACTORS

Ms. Jones is a low-level federal worker who does a combination of clerical and technical work. Actually, her job title is "systems analyst," although she says her work is mainly secretarial. She takes pride in what she knows, since she has been with her agency 16 years and worked her way up from strictly clerical duties to her present position. During that time she had moved into administration, but she says she suffered recent downward mobility as a result of the replacement of a political appointee for whom she had worked. Thus not only is her formal job title inconsistent with her acquired competence and everyday activities, but her job history has been uneven. This ambiguity about what she can do is the focus of her talk.

Mr. Smith is an upper-level administrator in a state agency, which he joined after working for five years on policy-oriented research at a private institute. Beginning as a research director, in six years he moved up to direct a broad category of programs. He says that he currently divides his time approximately evenly between administration and analysis. He speaks with gusto about his ability to learn quickly and his deliberate, continual efforts to find new challenges. In talking about his present position, he focuses on the organization's inability to provide him with new challenges. In fact, just prior to the interview, he had announced to his staff that he would be leaving in a few months to begin a new career as proprietor of a hunting lodge.

Every person, every case illustration, is unique. Neither of these people is representative, respectively, of all women or men, or of all administrative subordinates or superiors. However, the exaggerated themes of both dramas make it possible to see more clearly some bureaucrats' concerns about their autonomy as they confront the perceived constraints of collective regimens. These two have already been quoted in Chapter 3, both in connection with the organizational rumor mill. The first is the one who says that "I check for rumors as to what people perceive of me" but who, nevertheless, asserts, "I don't want to be recognized." The second is the worker who says that he pays attention to "the degree to which my name is bandied around the rumor mill. I think I am a more exciting person as a result."

Both are interested in their public performances. Indeed, both presentations have an ostentatiously dramatic quality. This interest in the performance begins with their comments about the interview itself.

When Ms. Jones gave me instructions about how I might recognize her when we met, she described herself simply by saying, "I am a blonde." She began the interview by saying that she had some things she wanted to say. At the end, she observed that I had made a transcript in shorthand, instead of using a tape recorder, and she expressed disappointment that some of what she had said might not have been preserved, and also, as a result, she had not gotten as much eye contact during the interview from me as she expected.

Mr. Smith remarked, on my guarantee of confidentiality, that he was a public person and preferred to be identified in whatever was written. He emphasized this by requesting I stay in the office if he received a phone call; his phone is a public phone, and what he has to say should be publicly known.

Each of the two uses a particular language and metaphor to represent the bureaucracy and his or her response to it. Jones's language is sexualized. Sex in the office and the tacit sexual content of relations between men and women are recurrent interests. Some of this concern may be the product of an idiosyncratic childhood; some may have been evoked by my being a man. Nevertheless, she makes insightful comments about the different treatment of men and women in a bureaucracy, and her perceptions of the sexual content of work relations are plausible. She does not obviously imagine sexual content where none exists; rather, she may be a more interested and faithful recorder of this realm than others. Importantly, her sexual interests seem less directed toward overt erotic activities than toward understanding how sexuality may provide a basis for establishing relations among people in an otherwise fragmented world.

Smith recently purchased a hunting lodge in northern Canada. Not surprisingly, he uses the languages of hunting and business to draw contrasts between the job he will soon start and work in the bureaucracy. Business language is useful in assessing public bureaucracy. The language of hunting, fishing, and nature is less commonly applied to bureaucracy; its use points to a number of artificialities in organizations. The languages of business and nature share a concern with risk. For Smith, both languages express facets of his personality. He observes that a high-school vocational interest test suggested he become either a real estate broker or a forester, expressing his entrepreneurial and natural interests.

THE PLAY'S THE THING: MS. JONES

Jones's drama centers on the problematic quality of organizational responsibility and meaning. Her drama is an actual play, or series of episodes, performed alongside formal organizational activities. In this drama she self-consciously plays a part:

> Because of the way I come off and the way I dress . . . I can deal effectively with men. When I was in the Director's Office, I was a brunette. I am now a blonde. Men are attracted to me as a blonde. I have undergone a personality change. There I was perceived as an overbearing bitch. I don't see myself as that. Now I am fun to be with.

This kind of performing may be culturally normal for some women, but what distinguishes Jones is her insistence on directing attention to the play. An important part of the motivation for this acting, as well as this emphasis on the dramatic artifice, is suggested by a comment she made early in the interview, where she observes,

> I am in a situation where I was raised to be an honest person. What I have learned is that in bureaucracy you can't operate that way.

She immediately continues with the example that

> This division has put some secretaries in training jobs and called them systems analysts. What they are doing is administrative work. That is what I am doing. I have gone to my branch chief and said, I want to learn the technical parts. They say, that is not what you are here for. I had some experience in the Director's Office. I am told that I know as much as many people who have been in systems a long time. I keep thinking that they know a lot, but they don't know I am doing administrative work.

Her workplace is full of such inconsistencies, or dishonesties. She points out that she works "in a section which is called Documentation; we do no documentation." She works for managers and administrators who "know nothing about managing and administration."

In a comment quoted earlier, she highlights incongruities between outdated job descriptions and workers' actual duties: "A lot of employees were in jobs which no longer existed." On its face, this statement is absurd; it is impossible to be in a job that does not exist. Clearly, what Jones means is that formal organizational policy is inconsistent with workers' actions. In this case, workers may take advantage of this job description inconsistency, but it is another case of organizational dishonesty. What makes this a potential problem is that, because officials can eventually enforce the organizational position, workers can be called to account for failing to meet the job requirements. Work that is not accounted for in this way is, in a sense, invisible.

As a result of cumulative experiences of organizational dishonesty, Jones says she "learned to question everything, even the obvious. Never assume, because it makes an ass of you and me" (assume = ass [u + me]). Because meaning in the organization is intrinsically doubtful, assuming carries the risk of being made an ass, a term which in both its animal and anal references implies shame and humiliation. Jones argues implicitly that it is not simply the structure of bureaucratic authority, but the inherent instability of organizational meaning, that confronts subordinates with the possibility of shame.

An anecdote about memo writing epitomizes organizational dishonesty and also illustrates Jones's response to it:

> For example, I got a memo, and I prepared a short memo response. I was told to rewrite it longer, and it became three pages, in order to look like a major project. I called one of the assistants in that office, and I told her not to judge me on the basis of the memo. But I can do it. Whatever they want and in the manner they want. You get a lot of that.

This anecdote reveals a great deal about her playacting. Meaning may be arbitrary, but there is an organizational hierarchy in which some can better enforce their meanings than others. Subordinates such as Jones must ultimately respond to the preferred meanings of those higher up. The second-to-last sentence expresses the feelings of shame that accompany such a command performance.

These feelings makes Jones's performance ambivalent. On the one hand, she says throughout that "the organization made me do it." It valued appearance more than substance, and it forced her to become an actress. However, while it required her to question the meaning of everything in order to avoid assumptions that would shame her later, her successful playacting shames her by violating her assumptions about her own autonomy. Shame is difficult to escape.

A subordinate may defend him- or herself against shame anxiety in several ways. One is to redefine shame-inducing actions (failures to perform appropriately) into guilt-inducing actions (offenses against moral standards), so that misdeeds may be remedied. Jones does not choose such a course, perhaps because she does not believe that she can get superiors to accept her definitions of situations or perhaps because she cannot accept the notion that she has acted inappropriately. Instead, she chooses two active strategies, in which she redefines the actions of her supervisors.[6]

First, she sets standards for their work and then finds their performances inadequate, and therefore, shameful. She does this with the managers-who-are-not-managers:

> I am clearly responsible to my section chief. I can get assignments from my division director, and I can say that I am working on this for him or her. Much of what I do is signed off perfunctorily. They know nothing about managing and administration.

In saying this, she asserts that her supervisors do not know how to do their jobs. Because she does know how to do their jobs, she sets standards for them. She does not tell them what they should be doing or how to do it, however, nor does she show them up. She just does their job for them, covertly, and the secret knowledge that comes from this performance forms the most supremely and satisfyingly shaming action she could take. She shames them, and yet they are so inept that they do not even understand that they are being shamed.[7]

Her actions illustrate ways in which many subordinates frequently enact dramas (Thompson, 1961). As organizational tasks become increasingly complex, managers with the authority to assign and evaluate do not always demonstrate the strength to solve the problems for which they are responsible, and subordinates with relevant competence may perform reparative dramas to save face for their superiors. Workers may engage in "double-talk" (Goffman, 1959), acting as if they were subordinate to competent administrators while covertly performing the administrators' responsibilities and expressing contempt for them. The covert action and message shame the superiors, but their pretensions to competence seem immoral because they destroy the meaning of responsibility. In this case, Jones decides to challenge them on moral grounds as well.

She signals this second strategy in her initial comments that bureaucracy does not permit her to act honestly, casting her relationship with the organization in moral terms and then indicting it for dishonesty—she will find her supervisors guilty. As with the determination of shame, the assessment of guilt is private. However, an assessment of guilt may be easier to impose because it can be concretely applied to specific actions rather than, as with shame, abstractly to an entire person.

The indictment of the organization for dishonesty makes disclaimed action the theme of Jones's drama. She acts but does not take responsibility for her actions, claiming "the organization made me do it." Her characterization of herself as "an honest person" and her condemnation of the organization for dishonesty are clear moral positions. Thus she puts on a play to show the organization that she recognizes its dishonesty, and she may nurture the hope of provoking individuals to respond to her honestly. Not only is the theme of the play disclaimed action, but also the fact of the play is disclaimed action: after all, it is only a play. Her play, where meaning is unreliable, is a response to the unreliability of meaning in the bureaucracy.[8] The question her performance poses for her is whether the play showing that meaning is unreliable is sufficient to endow her actions with meaning. If the (dramatic) play is to succeed as (psychological) play, she must eventually satisfy herself that she has forged meaning in a situation that lacked it.

"I Am Only Limited by the People Whom I Know"

Jones's play consists of a series of encounters with others. She finds the meaning of people's positions and relationships problematic; they are never what they are supposed to be. In her play, she attempts to show others what it is like when they cannot control meanings either.

Her description of the part she plays, quoted earlier, is the first disclaimer about her actions. She is not responsible for her actions because *she* is not acting. Instead, actions are being carried out by her demeanor, her dress, her hair, and her personality, each of which she distinguishes from herself. She is a stage for the play but not a character in it. Her lines immediately following the introductory statement repeatedly disclaim responsibility for actions:

> As a woman, I can get things done that men can't get done. Men basically want to help women. There are women in my section who can't get things. I have no problems with that. I do have problems with some women who have gotten into higher positions simply through physical use. I wouldn't do that. I have no problems using my God-given abilities, because men want that. Men don't get that same feeling helping fellow men.

These comments introduce the theme of sexual relations in bureaucratic organizations, but the disclaimers among the lines tend to deny Jones's participation in these relations. "As a woman," she can get things done; it is her gender, not she personally, that affects others. "Men basically want to help women"; because of their sex they respond spontaneously and predictably to women, Jones among them. She has problems with women who have gotten into higher positions "simply through physical use." "Use" is an incongruous word here, certainly passive, and again, it indicates that women do not participate actively in sexual relations.[9] Although Jones would not take part in overt sexual relations, she has no problems using her "God-given abilities, because men want that." Again, she is only a vehicle for others' actions; God called her to do what she does because men want it. "Men don't get that same feeling helping fellow men"; even the passions aroused in this play belong to others—in this case, men—not herself.

This play is about sexuality-which-is-not-sexuality. Why Jones creates the appearance of sexuality is suggested by the ways she attempts to use it. A prominent incident in her story involves a recent move of her office and efforts to have the new office space set up like the old one. She describes her efforts to influence the move:

> I am only limited by the people whom I know, unless I maintain a growing span of friends that includes the newcomers. I recognize that. I make a point of meeting people, getting them to know and like me. When the office next to us moved, they took one of the girl's partitions. It happened that I am friends with the branch chief. I said, do me a favor, et cetera. We need the partition. We need to separate smokers from nonsmokers. He said, by Monday you will have a partition. He is a new friend. I have new friends in the Director's Office. There is something in the human psyche that denotes authority. People assume that I have authority that I didn't know. There is something about my demeanor that denotes authority.

She refers again to this episode in summarizing her influence:

> Right now I have influence because of whom I know. I know people in high positions, and I can say, Hank, I have a problem, because they know and like me. My organization has had problems, and they know management will not support them, and they tell me to go up and mention it to So-and-so, and I have done it. I have influence because . . . for example, we are moving. We have different kinds of partitions. I am not in charge of the move, but we want partitions that can be anchored in the floor. The guy in charge told us we couldn't get them. And they told me, see what you can do. And because of my influence we are getting screw-into-the-floor partitions. I have no real authority, but I know everybody, many on a first-name basis. I have powers that I don't know.

The theme of sexuality-which-is-not-sexuality is suggested by the vague meaning of "knowing" in these lines. Jones "knows" people, including people in high positions. She is in some way familiar with men in her organization because men want to help women, and she needs their help for such mundane matters as getting office partitions. There is nothing extraordinary about such assistance. Part of her purpose in getting to know people is suggested by the terms in which she twice characterizes their response: people "know and like" her. Repeatedly, she refers to people in the organization as "friends." This dramatic organization of others may provide reassurance against anxiety aroused by unstable meaning by increasing the number of potential offenders (and defendants) (Fenichel, 1945). What she appears to be saying is that, whether or not there is any sexual content to the relations she establishes with men, she most values the relations, not the sexuality.

In short, she claims the relations but disclaims that sex is part of them. This disclaimer is explicit in the only lines in which she says she does *not* know something. She concludes the first statement by stating that she has "authority that I didn't know." This authority operates, unbeknownst to her, through her demeanor, to affect others. Indeed, this unknown authority, which may be exerted through the appearance of sexuality, acts for her only because others "assume" that she has it. Now others are the asses, victims of false assumptions. She has not claimed any intention to deceive others, as she believes she has been deceived herself, but the turnaround is implicitly just, not to say useful. The authority others give her clearly compensates for her lack of authority in the formal bureaucracy. She concludes the second statement by denying having any authority but acknowledging having "powers that I don't know." Again, the powers act on others independent of her volition. This disclaimer frees her of responsibility for actions and spares her from either the shame or guilt that might otherwise be associated with her actions. In these statements she lets the interviewer know that she knows that she knows-but-does-not-know what she is doing in her relations with men in the organization.[10]

The meaning of this drama is ambiguous. The message could be that in an organization in which meaning is unstable, friendly relations with others may provide the one dependable meaning. Social, even sexual, relations may substitute for logical relations. And yet she has disclaimed her actions in this play. She is not really responsible for those actions in which she apparently takes part. And, after all, it is only a play.

After the Play Is Over

At the end of the interview, Jones said that she did not want others to go through the experiences she had encountered in the bureaucracy. She hoped that whatever was written about her would warn others away from this danger. She added that she was beginning to develop her own private business. A number of her friends had begun to sell cosmetics, cleaning products, and the like in their own part- or full-time businesses. She felt that going into business for herself entailed financial risks, but she considered it a way out of the bureaucracy. For her it was a potentially meaningful activity.

THE CALL OF THE WILD: MR. SMITH

Smith succeeds in ways that Jones does not. Rising to a higher hierarchical position, perhaps competent in different ways, he succeeds in establishing clear meanings that serve his bureaucratic purposes. He can clarify responsibilities for others. He wants credit for what he does; he claims all his actions. And yet, in the end, as Jones, he complains that bureaucracy is ambiguous. However, Smith's complaint points to a second problematic aspect of bureaucracy. Jones charges that bureaucratic meaning is ambiguous: formal definitions of organizational phenomena are unstable and inconsistent with organizational practice. She characterizes the unreliability of meaning as dishonesty. In contrast, Smith charges that bureaucratic causality is ambiguous: there is no consistent or definite relationship between his actions and their bureaucratic consequences. He characterizes the unreliability of causality as "unreality." Repeatedly, he assesses the outcomes of his organizational actions as "ambiguous" and contrasts them with "more real" possibilities outside the organization. And he, as Jones, resolves his dissatisfaction by leaving the bureaucracy. His drama involves actual plans for the future.

Smith is himself subordinate to others. His subordinacy limits his influence—

his ability to cause others' actions—and yet the limitations on his influence are not all ambiguous. Describing his influence in the bureaucracy, he says,

> Influence, in my judgment, is the degree to which one exercises his independence. User status, your network of friends, or just like you are in charge and act like it. . . . Here, the exercise of influence is so contaminated with so many factors. Generally, there are three major components. Status in the organization, formal titles—if the Secretary says, 99 out of 100 times he gets something. Secondarily, when you cross operational lines but you are talking with people who are perceived to be on a higher level, you in no way can influence my day-to-day activities, but I know that when you send it to your boss, et cetera. Another thing—the informal network. It could be anywhere from outside the agency, the community, a politician: "This is the position I am taking; by the way, Senator Meathead said this." These are the major things.

In an important sense, causality in the hierarchy is not ambiguous. If the Secretary really can get what he wants 99 percent of the time, predictability is high. Similarly, if hierarchical superiors are normally able to influence hierarchical subordinates even when they do not have direct line relationships, causality, even if not certain, is not altogether arbitrary. In addition, there are regularities to patterns of informal influence. One might argue that Smith's causal ability is limited but unambiguous.

But Smith regards causality as more than simply predictability of outcomes. Outcomes, he insists, should have a reasonable—a meaningful—relationship to actions that precede them. In this sense, causality and meaning are inextricably linked. Thus he would consider causality to be ambiguous even in the context of the limitations on influence just described. These limitations largely reflect a hierarchical ordering of bureaucratic authority, which permits a limited range of action as a way of maximizing the regularity of workers' behavior and the predictability of organizational outcomes. Thus, beyond some point, no matter how much effort or what intentions a worker invests in a project, the consequences are likely to be more or less the same. There are limits to bureaucratic possibilities. Another way of describing these limits is to say that the relationship between actions and their consequences may have diminishing meaning. Metaphorically, one might contrast a game of pool on a conventional, finite pool table and a game on a vastly larger table. In both games the consequences of a shot are predictable, but the greater extent of the second table would allow effort to be reflected in more straightforward and conspicuous ways. The game on the first table is the conventional bureaucratic game. Smith wants the possibility of playing on the larger table, where possibilities are "more real."

Smith measures the "ambiguous" bureaucratic world against a model of a "more real" world. This model is embodied in his recently acquired hunting lodge. That project combines two "real" worlds. The first is that of the individual entrepreneur in the free market. Like Horatio Alger's waif, only smarter and less vulnerable, this individual formulates an idea for an investment, saves and invests the necessary money, awaits the interplay of opportunity and risk, and then confronts the outcome. If he has invested prudently, and if he has a modicum of luck, he prospers. If he has not thought the project through, perhaps if luck goes against him, he fails. The outcome is the measure of the man.

The second "real" world is that portrayed in American literary renderings of the pioneer life, in which man strives solitarily and gloriously against the forces of nature to establish his domain of civilization. Nature is worked and wooed to provide sustenance, but it can be unforgiving. Crops must be planted with care and attention to the vicissitudes of the seasons. Intelligence, care, and fortune separate those who survive from those who do not. Man's survival is the measure of his fitness.

Smith anticipates that outcomes and survival in the hunting lodge will be immediately measurable. If he has not invested thoughtfully, and if nature is not favorable, he can expect that his customer will complain to him: "I was here three days, the foot was rotten, it rained, and I didn't catch any fish." And he can expect that the customer may demand at least a partial refund, may not return, and may advise his friends not to patronize the lodge. However, if Smith works hard and wisely, the rewards can be substantial:

> There, I will be working during the season seven days a week, from five in the morning to nine at night. But I will see the results. People put $10,000 on the desk: "We have five people, we will stay here five days."

Indeed, with boldness and good fortune, within a few years Smith might amass both wealth and influence:

> It will really be dictated by the level of involvement that I want to pursue. Two to three years along, I will want to influence the planning of roads, airways, et cetera. That will be because I will own one of the largest lodges up there. "I am bringing in a half million dollars, you jokers, listen to me!" That turns heads.

For Smith, "reality" is measured by the immediacy of actions and their consequences. In contrast with the world of the hunting lodge, everything in the bureaucracy is mediated. Actions do not always have certain consequences, and the consequences of actions are not significantly related to the opportunities or risks confronted. For example, at the time of the interview, the most important project in Smith's work was his assignment to implement new federal legislation in the programs of his state agency. He confronted the challenge of maintaining services with declining federal support. He had worked long days and weekends for more than a year on both complicated technical analyses and intense political negotiations. In the end, he says, his primary organizational reward is compensatory time. But he is given further assignments that make it impossible to take the time off, and bureaucratic regulations specify that, if he does not take the time within two months, the time cannot be granted. This "reward" for success pales before the possibilities of the hunting lodge.

At the same time, some constituents of his programs, feeling aggrieved by his actions, threatened litigation. However, he observes, just as he is buffered from rewards for success, so, too, are his risks limited:

> The consequences are less severe on this job than they would be on that one [in the hunting lodge]. For example, if you want to sue me, you can sue the department, but you can never get me. But in that case [where he would own the hunting lodge] it is pretty clear that you can nail the corporate structure. To me it is more real.

Not only may he never be called to account for his actions before his critics, but, further, he is deprived of the opportunity of confronting and defeating his accusers.

The immediacy of the hunting lodge contrasts with the mediated "ambiguity" of the bureaucracy in several ways:

> Here I have autonomy, but [the hunting lodge] is the epitome of autonomy. This is important. Secondarily, working outdoors. You are working indoors in the bureaucracy. There you are out in the rain, the cold. Here you are in an office, overheated, overcooled. Thirdly, I enjoy working with my hands a lot. Another thing, the incentive system. I think the results are much more quickly realized [in private business]. I will know when I have made a mistake on the job. But secondary is the insulation from the economic realities of profit. In the lodge operations, I will not be. If you buy the wrong motors, you are down eight grand. Bureaucracies are a little slow to recognize or develop and instill an incentive system that rules financially or otherwise.

The theme in these remarks is that bureaucracy is unreal because it alienates man from his nature. First, Smith says, at best, he exercises only pseudoautonomy in the bureaucracy. Already, this represents a contrast with Jones and others who are hindered by shame from acting autonomously. Ambiguity of meaning is not a problem for him. In addition, he understands the exercise of bureaucratic authority and can assert it in ambiguous situations to carry out his responsibilities. However, he complains, no matter how definite he makes the meanings of bureaucratic phenomena, and no matter how authoritatively he acts, bureaucratic structures limit the range of consequences that follow his actions. In sharp contrast with Jones, who is wary of the perpetual dangers of being made a scapegoat for problems she cannot control, Smith complains that the bureaucracy offers no real risks equal to his abilities. Thus bureaucracy mediates between effort and outcome, either success or failure.

Second, bureaucracy mediates between people and nature. Bureaucratic offices are artificial environments over which an individual has no control. One reason Smith is moving to northern Canada, he says, is "geography—I am tired of the East Coast, the rotten weather." In an unusual choice of words, he characterizes that weather as "ambiguous," never definitely hot, cold, wet, or dry. The weather in northern Canada, in contrast, will never be uncertain, and he will be free to decide for himself how to respond to it. Confronting nature directly provides the real measure of the man.

Third, he charges, bureaucracy mediates between mind and body. Bureaucratic work is mental work, requiring physical exertion. Indeed, bureaucratic offices discourage physical activity. The extreme intellectualization of bureaucratic work further alienates people from their nature.

Finally, the faulty incentive system of bureaucracy has the effect of mediating between people and their potential. Results do not follow quickly from actions. There are no commensurate financial rewards for success. Nor does the bureaucracy offer even personal recognition. Moreover, despite pervasive anxiety about being shamed for inappropriate actions, even mistakes are haphazardly identified. As a result, people are deterred from confronting their potential.

"We're All in This Alone"

Jones produces a play-within-a-play in order to find relief from the anxiety of the larger bureaucratic world. Her play provides an escape in which she controls the plot and the outcomes are benign. At the same time, she plots a literal escape into some kind of private business. Smith plans an escape into the Canadian wilderness, but at the time of the interview his drama of the hunting lodge is largely fantasy. The ideal of the lodge offers a standard for measuring the frustrations of daily bureaucratic life and provides an escape from these frustrations. In order for the drama to succeed as play, Smith must convince himself that he will be able to control people and events in the wilderness as he cannot in the bureaucracy.

Both Jones and Smith seek to escape bureaucratic life in two ways: in imagination and in anticipated reality. Their motives are seemingly different. Jones finds her inability to control many bureaucratic meanings frustrating and shaming. Smith can control many bureaucratic meanings, and shame does not seem to bother him. He is frustrated, he says, because he cannot control bureaucratic causality, and within the bureaucracy cannot accomplish everything he believes he is capable of.

And yet, although Smith gives few overt signs of it, there seems to be something shaming about his bureaucratic life that, in the end, drives him out too. A suggestion of this shame is revealed in a comment contrasting the bureaucracy and the hunting lodge:

> The physical environment is important. I don't like a lot of people sharing with me what I want to do. This is not going to happen where we are.

He asserts his solitude—he does not want others to be involved in his projects. His final, summary statement in the interview affirms this position:

> I believe in hard work, because I am not a brilliant thinker. As the famed American sociologist Lily Tomlin said, "We're all in this alone."

This remark, simultaneously self-deprecating and self-respecting, humorous and serious, depicts solitude not simply as a natural condition, but also as an existentially inescapable situation. Society, even the cooperative activity in the bureaucracy, is ambiguously social: "We're all in this alone."

Other people, he is saying, are a problem. Other people interfere with his ambitions by opposing him or just by getting in his way. When he must depend on them to get things done, they are often unreliable. But other people also observe his failures. Smith has not formed notions of his capabilities in solitude. Through both actual and internal dialogues people from his past and the present help formulate his ambitions and join him in witnessing his successes and failures. He depends on these people for approval of his efforts, and yet he cannot assume that they will always find him adequate. In particular, they are often present to identify certain actions as his failures.[11]

Smith articulates Sartre's view that "Hell is—other people": specifically, shame is other people. Echoing the psychoanalysts, Sartre emphasizes the social constitution of shame:

> [A]lthough certain complex forms derived from shame can appear on the reflective plane, shame is not originally a phenomenon of reflection. . . . it is in its primary structure shame *before somebody*. . . . By the mere appearance of the Other, I am put in the position of passing judgment on myself as on an object, for it is an object that I appear to the Other. . . . Shame is by nature *recognition*. I recognize that I *am* as the Other sees me. (1943, rep. 1968, pp. 301–302)

Thus one may speculate that Smith's proposed journey to the hunting lodge is also an escape from shame in the bureaucracy. It is the shame inherent in any failure before others. In particular, however, Smith seeks to escape his failure to accomplish anything more than the mediated results permitted bureaucratic workers. In reality, probably most of Smith's imagined failures are invisible to his actual colleagues and friends, but—and this is what matters—he seems to believe otherwise. Even his significant organizational success does not spare him from the risks of such shame. And yet there is an ambiguity in the message of this drama, just as Jones's message is ambiguous. Jones suggests that anxiety engendered by organizational ambiguities might be assuaged by reliable social relations. At the same time, she disclaims the actions in her drama conveying this message and discusses her intentions to leave the bureaucracy to work as an individual entrepreneur. In apparent contrast, Smith suggests that the anxiety aroused by the consequences of organizational ambiguities might be avoided by seeking a solitary life. However, even as the entrepreneurial manager of the hunting lodge, he will work with family members, customers, suppliers, government agents, and others. The difference from the bureaucracy, he implies, is that this will be a society which, unlike the bureaucracy, he will be able to dominate and before whom he will feel no real shame. Still, as long as shame is inherent in social relations, it is uncertain whether the hunting lodge can bring the peace he imagines.

THE MESSAGE ABOUT BUREAUCRACY IN THE DRAMAS

The two dramas provide escape. Each creates an alternative world in which bureaucratic threats are contained and damages are repaired. They are also dramas of compensation. Workers have produced the plays to secure the responsibility and authority, the meaning and causality, which bureaucratic predicaments withhold.

Jones responds to the predicament of ambiguous responsibility by creating an alternative world in which she can play with ambiguity. Some of the ambiguity in her drama's message reflects the differences in ways in which she experiments. If she wishes, she can influence others and yet obscure her role by generating counterambiguity. For example, she colludes with inept administrators by doing their work competently, while they receive the credit. Often she acts and yet disclaims any role in her actions: it is not she, but her personality, her demeanor, her hair, or her clothes which have acted. And yet, if she chooses, she can establish unambiguous responsibility, along with commensurate authority, and act competently. After all her disclaimers, she really is responsible for getting the partitions for the new office, and she takes credit for it. In this dramatic realm she can get recognition for her performance.

Smith's drama similarly holds out the promise of unambiguous meanings and responsibility but, in addition, offers a remedy for the dangers of ambiguous authority. In his dramatic world he experiments with causality. Although the drama is ambiguously both a realistic forecast and a wishful anticipation, causal relations in the drama itself are certain. With little ambivalence he imagines situations in which he has definite authority and can make important things happen. Customers and government agents alike respect him. This unambiguous authority, in combination with unambiguous responsibility, enables him to perform competently and be recognized for success.

Jones's drama responds to the predicament of autonomous authority by experimenting with the clarity of authority. As a playwright, she understands authority; it is transparent to her. Sometimes she chooses to reveal what she knows, as in her phone call to the woman to whom she sent the expanded three-page memorandum or in parts of the interview. At these times she imbues bureaucratic relations with such honesty that false assumptions are unlikely. At other times she prefers to imitate the autonomy of her superiors by influencing others while hiding in seductiveness. In both ways she can act without being shamed; in the latter, she gains the control others' autonomy has denied her.

Smith's drama responds to this predicament by experimenting with authority. Unwaveringly, he creates autonomy for himself. He will be absolute ruler of the wilderness, and he will humiliate government officials by making undeniable demands for roads or airways. Because they will need his business, while he will be indifferent to them, they will comply meekly. In this dramatic world he is subordinate to no one, dependent on no one, and no one can shame him.

Jones's moral indictment of the bureaucracy for dishonesty and Smith's charge that the bureaucracy is unreal and unnatural speak for the pain and anger which these predicaments may evoke. The feeling aroused may be measured in both dramas' vengeful wishes to humiliate others. The dramas may be seen as efforts to undo the bureaucratic relationships that challenge workers' competence and recognition and threaten their security. By creating new organizational settings, these dramas nullify workers' prior commitments to the organizations causing such pain. At the same time, these dramas take revenge on the bureaucracy by denying its claims and significance. Unfortunately, in undoing the formal bureaucracy, these dramas lead workers away from work and problem solving.

The Causes of Escape

To what degree does bureaucracy itself contribute to workers' desire to escape? Are workers possibly projecting their personal dissatisfactions onto the organization and blaming it for their inner conflicts? Is the interest in escape anything other than a reflection of individual personality? If so, what part of the motivation to escape comes specifically from bureaucratic conditions, and what part from social life generally?

Each of the dramas points to underlying personality characteristics. Jones's pseudosexuality, her continual disclaimers of action, as well as her dramaturgical interests, hint at the operations of a hysterical character (see, for instance, Fenichel, 1945). In addition, she is not quite candid in saying that the organization made her do it—namely, take up dishonesty. At the end of the interview she briefly mentioned her stepfather. He continually denied her things she wanted. For example, he refused to let her go to dances with friends. She reacted to this frustration by telling him she didn't really want to go anyway. This suggests that she took up dishonesty in childhood, at least partly as a way of protecting herself from being shamed by an autonomous, possibly cruel stepfather. One might speculate that, as a result of these experiences, she tends to see work relations in similar ways.

Smith's drama offers evidence of a narcissistic personality. In a peculiar way he simultaneously expresses a grandiose view of his potential and feels insecure. He believes he is capable of great things—challenging people as well as nature—but he dislikes having others around. He feels ashamed before them; he is not yet the big man he wants to be. In his plans he compensates for his insecurity. He conjures up a majestic environment to mirror his grandiose self. In the end, everything, from the forests to government agents, will give in to his power. In the world he anticipates no one is as important as he is; he depends on no one; he is autonomous, self-sufficient (see, for instance, Kernberg, 1975; Kohut, 1971). Although Smith said little about his past, this way of thinking is probably linked with his earlier experiences.[12]

The important question is whether these dramas and the intentions they reveal are only expressions of personality, with no relation to the work situation. There is no definitive answer for this question, but the dramas offer evidence of bureaucratic influence. First, the dramatic texts themselves link the workers' interests in escape to the organization. Jones calls attention to the "dishonest" aspects of her job as a motivation for her play. She continually refers to contrasts between her drama and the bureaucracy as a way of showing, first, that the two are related and, second, that her dramatic world represents an improvement over the organization. Smith explains why he wants to leave by making repeated comparisons between his bureaucratic job and entrepreneurial management in a natural setting.

Second, even though not all bureaucrats describe daydreams, sidegames, or plans to resign, these dramas are consistent with what many workers say about bureaucratic predicaments. The dramas describe a similar psychological structure and offer plausible responses to it. Jones's and Smith's personalities are not obviously pathological. Their pasts surely color their present perceptions, but they also provide special sensitivities. Jones may be particularly sensitive to inconsistencies in bureaucratic meanings and role relationships. Smith may be especially sensitive to limitations on accomplishment. Although their presentations may be flamboyant, what they say is neither bizarre nor unreasonable. Their dramas of escape symbolize common wishes to escape.

The Consequences of Escape from the Bureaucracy

What does it mean to say that workers "escape" from the bureaucracy? How tangible is this escape? And what are the consequences in the performance of work? Some escape is literal. Smith, who carried out high-level administrative work, has resigned and made definite plans to move to northern Canada. Jones, who has been with her agency for 16 years, weighs the possibility of leaving the bureaucracy to set up an independent business.

Workers may remain in their organization physically but escape in ways that divert energy from their formal tasks. Jones's drama involves tangible activities with a number of other workers in relations that are not directly linked to formal organizational responsibilities.[13] Her plot may be unique, but her intentions and activities are not. Other workers create or participate in similar dramas, with significant consequences for organizational accomplishment. For one, simultaneous performance in formal organizational roles and in dramas of escape requires considerable effort, as it entails being in two "places" at once. The dramas inevitably draw energy from the formal roles. Tangibly, workers may be unable to give undivided time or attention to the relationships, research, or documents their formal roles require. In addition, the stress of switching and balancing roles may diminish the quality of work. In particular, insofar as the dramatic roles bring greater satisfaction and security than bureaucratic roles, workers will want to stay in the dramas, and disengagement may require an especial expenditure of energy.

Problems in the dramatic worlds, where workers may find it easier both to exert power and to collaborate, may be more compelling and seem more capable of solution than problems in the bureaucracy. As a result, workers may spend considerable time solving problems, but the problems may be unrelated to organizational goals or formal work assignments.

NOTES

1. See Waelder (1933) for a presentation of the view that play generally is an effort to repeat and master traumatic relations with the environment.

2. For an analysis specifically of children's grappling with separation from their mother, see Bowlby (1973) and Mahler, Pine, and Bergman (1975).

3. For a discussion of differences between children's play and adults' play, see Erikson (1977) and Peller (1954).

4. Schafer (1968) describes daydreaming in these general terms, although it should be clear that the dramas discussed here are often more active than mere daydreaming.

5. In these terms, it would be inaccurate to say, for example, that the psychological structure of bureaucracy "causes" the actions described in this book. Instead, organizational members consistently give bureaucratic social relations meanings that appear to justify the actions they report. Although this relationship corresponds to conventional meanings of "causality" in the social world, it should be clear that there is no irrevocable or Newtonian relationship between a distinct social structure and individual actions. If workers changed their perceptions of relations with others, they might modify their actions or descriptions of them as well. For an examination of this issue with regard to the language of psychoanalysis, see Schafer (1976).

6. Taking active steps to control something passively suffered is, as Freud (1928; rep. 1959) observed, a fundamental principle of psychological defenses. Waelder (1933) emphasizes that this conversion of passivity into activity is basic to play, of which Jones's drama is an example. What is an active strategy in specific circumstances, in contrast with what is passive, depends on each individual's conscious and unconscious interpretations of past experiences.

7. For a powerful literary comparison to the logic of Jones's statement, see Kafka's *Letter to His Father* (1919; rep. 1966).

8. The conspicuousness of her playacting recalls Freud's (1930; rep. 1965) observation that children's absurd games or fantasies are often aimed at ridiculing adults: "Since you lie to me in your way, I shall lie to you in mine."

9. Her reference to physical "use" is interpreted here to mean that she believes that some women submit their bodies to physical use by men in return for advancement. It is possible that she means something more active—that women deliberately use their bodies to negotiate promotion. However, although such "use" may appear to be voluntary, the purpose of the relations is not sexual gratification so much as organizational reward. In the end, women's bodies are used as tokens for bargaining in a definitely unequal power relationship.

10. Her disclaimed action is analogous to planners' denials of intentions to exercise power. Many planners say that they want to have influence but do not want to organize overtly to exercise power. When confronted with the necessity of using it to implement goals, they tend to deny interest in the results. A common concern of Jones and many planners is anxiety about the risk and vulnerabilities of confrontation when organizing.

11. Smith's assumptions about others' expectations of him represent what has previously been described as an ego ideal. Whether others actually said they expect certain things of him is less important than that he believes he must live up to these high standards.

12. Although these clinical sketches are consistent with the interview data, it should be clear that the relatively brief, nonclinical interviews are insufficient for making deep personality assessments.

13. The human relations school of organization (see Argyris, 1960; Likert, 1961; McGregor, 1960; Roethlisberger & Dickson, 1939; Schein, 1965) offers a somewhat different interpretation of such activities. In that view, these activities are linked with formal organizational responsibilities in the sense that, if they are not carried out, then workers are unlikely to invest themselves in organizational tasks. That interpretation puts a higher value on the interests of the organization than on those of the workers but does not really contradict the empirical observations in this chapter.

6

A Technical Defense in Practice

"But 'glory' doesn't mean 'a nice knockdown argument,'" Alice objected.

"When *I* use a word," Humpty Dumpty said in rather a scornful tone, "it means just what I choose it to mean—neither more nor less."

"The question is," said Alice, "whether you *can* make words mean so many different things."

"The question is," said Humpty Dumpty, "which is to be master—that's all." (Lewis Carroll, *Through the Looking Glass*)

Even though many bureaucratic workers think of escape, unless they literally leave the bureaucracy, they must spend much of their time on their formal organizational tasks. This chapter examines how workers respond to expectations that they help solve organizational problems.

Planners, again, illustrate workers' dilemmas and responses. As professionals, planners are expected to recommend solutions for a variety of social problems. Often they have little control over the schedule for their work, and a statute or regulation may prescribe a fixed time for the delivery of a plan, proposal, or recommendation. Even if the problems are complex and ambiguous, planners are expected to suggest clearcut solutions for them. In these respects, they confront expectations shared by many bureaucratic workers.

The planning process, then, leads toward an unambiguous solution for a problem. Insofar as practitioners feel anxious about asserting any definite position, they are likely to feel increasingly anxious as they move through the progressive steps in this process. Expected to do just what arouses anxiety, many planners attempt to control the planning process so as to avoid shame or guilt. These defenses can be inferred from the ways that their actual problem-solving behavior diverges from the model they espouse.

AN ESPOUSED MODEL OF PROBLEM SOLVING[1]

When asked how they go about solving problems, most planners say they follow a model of open-ended search and analysis, such as described by this county planner who portrays planning as competence

> in analyzing design and . . . in formulating ideas . . . the ability to evaluate alternatives; ability to identify different threads or facets of a problem, to identify all the various forces that are at work on a particular conflict, and to come up with a scheme which resolves these conflicts.

Because of the ambiguity of planning situations, formulating a problem appropriately is at least as important as solving it. Another planner emphasizes this point:

> I think the thing probably is in synthesis. I am not a data freak. I enjoy getting together all the information because it is necessary. But the challenge is to say, what does it all mean? Which I would call the synthesis phase. That is what I can contribute the most.

Planners must always assess problems in terms of the future. Which current problems may go away without intervention, and what new ones will supersede them? What course of action may be most effective in dealing with future problems? Detachment from present circumstances permits a broader perspective on current problems, and a local planner emphasizes his

> ability to remove myself from the day and look at the future to get a decent reading on what is happening . . . in order to have an air of predictability which you can be relatively sure about.

These statements compare favorably with textbook norms emphasizing an extensive, inclusive collection of information. The ideal planning process is said to require "an *exhaustive survey* [sic] of all relevant alternatives, since an unstudied alternative may be the optimal one" (Etzioni, 1968, p. 264). Consistently, the second planner has said that he gets together "all the information because it is necessary." In particular, the ideal planner should collect sufficient information that "the decision-maker *considers all of the alternatives* (courses of action) open to him [and] *identifies and evaluates all the consequences which would follow* from the adoption of each alternative" (Banfield, 1955, p. 314; emphasis mine). Consistently the first planner has reported that he identifies all the various forces at work on a "particular conflict" and develops "a scheme which resolves these conflicts."

This espoused model emphasizes an extensive search for information. And although these planners say little about the sources of information, a comprehensive search will put them in contact with many people. In addition, the process of recommending proposals necessitates extensive contact with others. To the degree that a search for possible solutions is widespread, any particular recommendation is likely to be new, even potentially threatening, to many people. The feasibility of any proposal depends on its acceptance by those who could implement it. Thus finding a solution may entail persuading people that a particular course of action could solve a problem.

TECHNICAL SPECIALIZATION AS DEFENSE

Cognitively, these principles for solving problems are reasonable.[2] However, insofar as they call for planners to confront others, including superordinates, these norms may arouse shame and guilt anxiety. Because problems are supposed to be solved, direct retreat is impossible. Instead, many[3] planners seek to defend themselves by promoting a strong offense. They attempt to claim authority that is independent of either the bureaucratic organization or the political arena—the authority of technical expertise.

Such a claim is typical of the professions. Professionals are said to differ from laypersons in possessing a specialized expertise about certain important problems. In addition, professionals adhere to a relevant ethical code, which ensures that their knowledge will be used for good. For these reasons they may be permitted to work relatively free from surveillance or regulation.[4]

In invoking these conventional professional claims, planners might contend that they know better than others—including their hierarchical superiors, members of the public, and elected officials—how social problems are to be analyzed and solved. Whoever submits a problem to planners should accept their recommendations as the word of experts. Such an argument would shift the power relations in bureaucratic organizations by invoking an independent source of authority for planners.[5] Such a claim, if persuasive, would grant planners, despite their hierarchical subordinacy, decision-making authority challenging the autonomy of their superordinates. It would authorize planners to evaluate their own work and, by implication, that of their supervisors as well. Planners might reverse the relationship in which supervisors' autonomy continually threatens to shame subordinates. Professional claims thus are an antidote for shame anxiety.

In American society professional status and autonomy are most readily bestowed on practitioners who demonstrate a specialized competence in some "technical" domain. Unfortunately, the peculiarities of the organizational and societal problems on which planners, as well as administrators, analysts, and many other bureaucrats, work make such claims difficult. Problems transcend conventional intellectual disciplines, and a general knowledge is necessary both to draw workable bounds around problems and to recognize potentially effective interventions. The variety and complexity of these problems place a premium on the skills of a generalist, who can see similarities among problems that appear superficially different and who can apply perspectives from one field in another, uncustomary context. Thus effectiveness in social problem solving depends on generalism.[6]

Planners may be more competent at this work than laypersons, but the competence is an experienced generalism, rather than a narrow, deep specialization. Accurately defining their real skills, then, would not satisfy culturally prescribed expectations of professionals and would limit any chances for autonomy.[7] Thus an acknowledgment of generalism holds planners open to continuing risks of shame for anticipated assessment of failure by hierarchical superiors.

The incongruity between planners' expertise and cultural expectations of professionals contributes to planners' doubts that they have a legitimately defined area of competence. At the same time workers who are incapacitated by doubt do not assert themselves or expect success and, consequently, avoid the anxiety of anticipated failure and shame. However, if doubt enables planners to gain these unconscious aims, it also defeats conscious interests in competence, assertion, and authority. Hence, despite the weakness of claims of specialization, planners may still seek to argue that their work is technical. The complex feelings associated with these issues are reflected in the ambivalence with which planners discuss their technical skills, including the intensity with which many planners attempt to resolve the ambivalence by insisting that, in the end, they are technicians.

Technical Is as Technical Does

When planners are asked what expertise distinguishes them from laypersons, many refer to technical skills. For a relatively small number of planners, "technical" skills have a clear and specific meaning, as in the following observation about technical aspects. The public doesn't have expertise and time. The mechanics: what type of building material, actual siting considerations. For example, how many personnel in a medical clinic.

However, many planners doubt that they have definite, exclusive technical skills. Some recognize that technical expertise is important in planning but regard themselves as generalists. In this respect, they may be similar to their bureaucratic and public clients:

> It is difficult for citizens sometimes to understand or be involved in the technical aspects of what we do. But I am not sure they need to. I don't understand all the technical aspects.

Many planners find it difficult even to share vicariously in the skills of others who are specialists, and as a result, cannot distinguish between themselves and laypersons:

> You see different individuals being able to do different things, and I don't know that there are particular things about the steps that you go through in the process that any citizen or group of citizens that are generally knowledgeable couldn't do an adequate job. I don't think that by definition there is any skill which may not be available in the community. I think that planners put together many skills at once. I don't think there is anything profound about this, so that citizens could participate in any phase of the process. A different question is whether citizens have the interest or the time.

But such doubt leaves planners in the dilemma just described. The discomfort many planners experience in relations with the public is expressed in the following comments, where the planners justify their autonomy by pointing to citizens' irrationality:

> Citizens' involvement is very, very important. I have attended public hearings. I have the feeling that people here have no confidence in technical people, technical knowledge, and they question technical proposals. It's good, but if you waste time responding to nonsense, you get slowed down.

> In theory, in school, you hear that's great, citizen participation. But they're a pain in the ass. You have some people who are professional citizens, who have a chip on their shoulder. But that is an elitist position. I just wish they'd trust me. I don't enjoy working with citizens at all.

These planners are not simply ambivalent about working with citizens; they feel vulnerable to harm from them. The first planner, expressing qualified support for citizen participation, complains that citizens "have no confidence in technical people"—namely, planners. The second planner echoes these sentiments. Too often, citizens may slow down work. Much worse, citizens are "a pain in the ass." Interpreted literally, "a pain in the ass" is a humiliating violation of both body and dignity. It is the extreme transgression of any possible boundaries between planners and members of the public. The specific emotion to which "a pain in the ass" refers is shame anxiety—apprehension about being found out not to have esoteric skills or knowledge. If this happens, then members of the public may legitimately oversee planners' actions—or worse, may ignore planners altogether.

In the face of such anxiety, it is perhaps not surprising that a number of planners talk tautologically about their technical skills. Planners possess "technical" skills, which, by definition, laypersons do not have. For example, "The more technical it is, the less possible it is for the ordinary citizen to get involved." "To get involved in all the technical things is beyond the ability of the average citizen." The implied social meaning of these statements is clear: they justify excluding others from decision making. This social function is illustrated in the following statements, where planners clearly link their claims to technical skills, with professional autonomy:

> [What citizens are least capable of doing is] the real analytic task and probably the real design task. . . . This is what you as a professional are supposedly trained for. These are your unique abilities, what the average citizen does not have.

> [Citizens are least competent at] technical analysis, but I have often found an interesting insight in nontechnical analysis of technical analysis. I have seen some intelligent analysis of grandiose schemes which planners have developed. But also the technical area is still an area of the planner's expertise. While the citizen should be cautious of planning, he should still be aware that the planner is an expert.

Many planners are certainly more competent than many citizens, and many planners wield expertise which might be considered specialized and technical. However, the "supposedly" in the first comment and the "interesting insight" and "intelligent analysis" in the second express doubt about this. In response, some planners may emphasize their "technical" competence primarily to establish social distance from others, including citizen groups and administrative superiors. The "but also" and "still" in the second statement, echoing an implicit "but" in the first, insist on this demarcation.

AN IMPLICIT MODEL OF PROBLEM SOLVING[8]

Specialization may improve the problem-solving process by increasing the information collected and adding sophistication to its analysis. However, insofar as practitioners invoke technical expertise as a defense against collaboration with others, then less information may be collected, and it may be subjected to an unsystematically limited analysis.

Planners confront this dilemma in relations with the public. Although citizens are sources of information, they are also political actors who threaten planners' ability to control their relations with bureaucratic superiors. This latter concern leads many planners to limit citizen influence by depreciating what they say. Many planners ignore public and bureaucratic clients by acting as if they were self-employed entrepreneurs. These defenses may be seen in the following analyses of planners' attitudes toward citizen participation and their evaluations of success.[9]

What Do Laypersons Know? The Meaninglessness of Political Information

Citizen participation is legally required for many planning programs. In addition, in many agencies it has become an institutional norm, and in many communities it is a public expectation. Planners themselves offer two justifications for citizen participation. One is political—if citizens are involved in discussions of public problems, they will give legitimacy and support to any specific proposed solutions. A second claim is intellectual—citizens have information that planners do not. If planners are interested in comprehensive information collection, then some process of public consultation is necessary.

However, although most planners concede that laypersons have useful information, they tend to define laypersons' expertise in such a way that their ideas may be discounted. Most planners believe that the citizens' primary contribution to problem solving is parochial information. Residents have an intimate knowledge of conditions in their neighborhood. They understand the quality of local life in ways that formally collected quantitative data may not reveal. Accordingly, three-fourths of planners interviewed regard citizen participation as important in the beginning of problem solving, when problems are identified. The following is a typical description of what residents may contribute:

> A well-informed citizenry . . . can be helpful in helping the professional understand a neighborhood, can point out these things which from a planner's experience we might not know. . . . There may be planners who are looking at data who may not understand what they are looking at with the data: for example, a statistical description of housing prices. There is something which is qualitative which is not apparent in the numbers. There is a need for interpretation of quantitative data. Neighborhood groups can provide guidance on customs of the neighborhood, et cetera.

Once community members have finished describing local conditions, however, many planners want to exclude them from further participation. Emphasizing professional skills required, planners contend that the public is cognitively ill-suited for the core work. Once the problem is defined, it must be interpreted in order to formulate goals for action; only half of the planners interviewed believe that citizens have the requisite reflective ability. Once goals have been established, alternative strategies must be developed and evaluated for their feasibility; at this point only one-third of planners believe that citizens can abstractly analyze the choices involved.

In short, members of the public have concrete interests, whereas planning requires an understanding of abstract principles.[10] Thus laypersons cannot see the large picture. Spatially, unlike planners, they cannot understand how conditions in their own neighborhoods are related to conditions in other neighborhoods and how they all affect and respond to metropolitan or regional conditions. Temporally, unlike planners, citizens cannot understand what is the likely course of immediate conditions over a long period of time.

Couched in terms of cognitive ability and style, these criticisms imply that citizens do not know what they are talking about. Planners can understand better than community members what the members of the public mean, or should mean. The following characterizations are typical:

> They are least able to—and this is one of the weaknesses of citizen participation—they are at least able to see the long-term comprehensive benefits of a project. They are most concerned with how it affects one area directly.

> [They are least capable of having insight on] anything which has long-range implications. I don't think a community, because it has self-preservation interests, can make proper decisions for the interests of a larger community, the whole metropolitan area.

> . . . I think the planner is more aware of the larger implications than the neighborhood itself.

These "larger implications" represent the public interest, which planners believe they are accountable for and are most competent to identify. In contrast, laypersons' cognitive narrowness leads them to act only on behalf of what they can see: private "self-preservation interests." In general, this characterization describes citizens as incapable of accurately understanding what social phenomena mean. Only planners, because of their cognitive ability, can understand social meanings, and thus should anticipate the necessity of overruling community members. In particular, this cognitive criticism of community members undermines statements about self-interests. Planners who take the position that the only legitimate interest is the public interest (however that is defined) implicitly reject any assertions of community interests as cognitive errors rather than as acceptable political claims.[11]

In Humpty Dumpty-like fashion ("'When *I* use a word,' Humpty Dumpty said, in rather a scornful tone, 'it means just what I choose it to mean—neither more nor less.'" [Carroll, 1871; rep. 1931, p. 238]), most planners assert that their concern for the public interest is "rational," whereas they tend to characterize attention to other interests as "political."[12] Such a label is accurate insofar as it means that citizen actions are self-interested and organized. However, in this case "political" is used pejoratively to mean contravening the dictates of rational analysis. The "political" is irrational. If reason is the essence of problem solving, then laypersons, who tend to be "political," should be excluded from planning.[13] This attitude appears in the following statements.

> [In my experience, citizens] were not interested in planning decisions. They were fascinated by the political process. But they couldn't translate their experience into decisions about the parcel of land. People identify issues which you need to respond to. The professionals are accountable [for analysis and recommendations]. Citizens couldn't recognize alternatives or weigh alternatives in a large perspective. Citizen consultants get bored at that level.

> I don't think they have the perspective of all the viewpoints that have to be taken into account. I don't think they understand how decisions are made. I think it is the role of the planner to meld all these perspectives and account for them but make the final choice.

The exclusion of "political" information is a counterpart to defensive claims of "technical" expertise. Insofar as politics is considered to be everyone's business—and to the degree that no one, with the possible exception of full-time politicians, is considered a political authority—the admissibility of political matters in the planning process would challenge planners' claims to superior expertise in problem solving. Hence planners may simultaneously claim jurisdiction over a broad range of problems and assert authority in defining the information relevant to solving those problems.

This tacit model of problem solving revealed in planners' comments differs sharply from the espoused ideal. Despite claims to comprehensive information, collection, this process depends heavily on planners' judgments about what information is meaningful. Their claim to technical specialization justifies excluding generally informed laypersons. At the same time, planners' claim to a broad view of problems justifies excluding laypersons for being too narrowly informed and interested.

The exclusion of political information protects not only planners' intellectual control over the planning process, but their social control as well. For if political information were considered germane to solving problems, then planners would have to meet more often with members of the public—and in settings often more intensely argumentative than deliberately reasonable. Such confrontation would render planners vulnerable to challenge. People might discover planners to be generalists who work by rules of thumb, rather than specialists who follow technical algorithms. Planners' assertion of authority over meanings defends them against such a discovery and the attendant anxiety.

Thus claims of technical expertise operate paradoxically. Planners appear to assert that problems may be subjected to well-defined analytic methods that will produce specific unambiguous solutions. However, the main effect of technical claims is to tell others that their perceptions of problems are inadequate. This use of technical claims is an example of counterambiguity, through which planners seek to convince others that situations are different than they seem. With both bureaucratic superiors and public clients, planners invoke responsibility-without-responsibility. In different contexts, agency directors and community groups may believe that planning staff members are accountable to them. However, planners may interpret their expertise to mean that they follow only their interpretations of rationality, thus construing their accountability to give them autonomy. When justifying their recommendations, they may say that they personally make no decisions, but only do what rationality permits. In particular, many imply that rationality requires them to overlook or reinterpret anything elected officials or citizen groups think and say about social conditions. As a result, others may find that their assumptions about both urban problems and relations with the planner seem wrong, while the planner apparently has no discretion in ignoring or overriding their assumptions. Thus invoking technical expertise, a planner may simultaneously escape criticism for handling a problem as a superior or client defines it and gain control over the problem-solving process. This turnabout reduces the planner's risks of being shamed for any analysis or proposals.

The Entrepreneurial Problem Solver

Planners work for clients, who give them assignments. Elected officials, private developers, or community groups may ask planners for studies or proposals. In addition, public agency planners are responsible for serving certain statutorily defined public clients or constituencies, such as people who need housing.[14] Although planners occasionally formulate their own projects, this practice is exceptional, and success still requires the adoption of the project by some potential clients. Planners essentially work on problems for others, who ask planners for advice but want to choose the solutions themselves. Use of the planners' ideas depends on their acceptability to the clients receiving them. This dependency, as has been noted, is potentially shameful. It is also potentially frustrating, insofar as many intellectual efforts may never lead directly to concrete results.

A number of planners defend themselves against the anxiety and frustration by thinking of themselves as independent entrepreneurs. This aspiration is supported by dominant images of professionals, who are normally independent, self-employed practitioners. Accordingly, despite claims to represent a public interest, many planners insist on working privately, preferring the undistracted solitude of their offices to the public give-and-take of community meetings and consultations. Many are evidently more concerned with the internal rationality of their analyses than with their accountability to the public. Many see themselves as guardians of reason and the public interest.

Planners describe this entrepreneurship in the ways they measure their success. Many say they have difficulty assessing their work. A primary source of difficulty is the frequent divergence between technical and political requirements for solutions of problems. Planners often say they understand the former but the latter are mysterious. As one says, "Technically you know what you did. [But you] don't know the significance ever." In self-defense some planners make two separate assessments of their work: one of its technical quality, about which they are confident, and another of its political or organizational usefulness, about which they understand much less. An experienced planner, quoted earlier, describes this separation:

> In terms of whether or not the work I do is *right*, I don't have all that many qualms about. Whether or not it is *useful* is where I am on softer grounds where assessment is concerned. [For example, I don't know whether my last major project] is related to improving life in the city.

Planners seek indicators of their effectiveness. More than two-thirds of planning administrators interviewed say that, in measuring their success, they look for implementation or use of their work, praise from clients or constitutents, or both. Clearly, administrative success depends on concrete accomplishments and public appreciation. In addition, administrative positions offer authority for promoting recommendations, and expectations of results are realistic.

In sharp contrast, subordinate staff members say little about either implementation or public appreciation. These are matters of "usefulness." Instead, they focus on the "rightness" of their work in the eyes of other technicians. Approximately two-thirds of the subordinates interviewed emphasize merely praise from other agency workers as the indicator of their effectiveness. One planner puts the matter simply: "What is primary for me is whether my work meets criticism, praise, compliments." At the same time, the scarcity of praise both enhances its value and encourages planners to hold modest expectations even in this regard, as another planner quoted earlier complains:

> [I look for] no editing [of what I write]. . . . There is not that much backslapping and congratulations. When I have done something decent, I get some congratulatory remarks from colleagues. I do not need that all the time, but now and then it is nice. Often here you feel you are being taken for granted.

In this view, citizens are inconsequential. Moreover, administrators, while always present, are not to be counted on for appreciation or even recognition. This encourages some defiant self-reliance:

> [I am concerned about] my satisfaction, period. . . . Sometimes when I am doing bits of work, I am able to say it is a good piece of work, whether they like it or not.

Even though these planners are subordinate to bureaucratic administrators and accountable to public clients, they tend to portray themselves as private entrepreneurs. They ply their trade with craftsmanlike care and anticipate recognition of their skills from other craftsmen.

The entrepreneurial formulation of the problem-solver's role completes the tacitly defensive model of problem solving. The "technical" rejection of "political" information justifies disregarding what lay persons may say. The entrepreneurial role relieves planners of accountability to others for either considering their information or producing recommendations that take their concerns into account.

Both actions may be considered strategies for undoing the bureaucratic setting that engenders anxiety and frustrates aims for a psychological contract. When planners emphasize their technical competence, they deny the need to be responsive to others. When planners think in terms of an entrepreneurial role, they deny the reality of the bureaucracy and organizational accountability altogether.

THE POWER OF TECHNICAL RATIONALITY

Planners' implict model of problem solving apparently corresponds to a Type IV power orientation. The orientation is suprapersonal, in that an other is both the source and object of power. In this orientation someone is moved by high principles to serve or influence others. The principles are the impetus for the actions, the one who acts only serves those principles, and the consequences of the actions adhere to the principles rather than to the actor. Many planners contend that they are moved by the dictates of rationality to make recommendations serving the public interest.

The question has arisen, how, if the four power orientations represent a developmental sequence, planners apparently stuck in a Type II orientation (strengthening oneself), deterred from learning a Type III orientation (influencing others), nevertheless can seemingly express also a Type IV orientation (serving higher principles or group interests to influence others). Chapter 4 suggested two possible explanations.[15] First, some planners may be capable of all four power orientations but may choose to work in the second and fourth as most consistent with professional norms. Alternatively, some planners may be competent in only the first (being strengthened by others) and second orientations but may employ a pseudo-Type IV orientation in service of the second orientation. Thus claims of serving the principles of rationality might be invoked as a means of bolstering interests in strengthening and controlling a role in collecting and analyzing information. It is impossible to know without much more information, but probably some significant proportion of planners' actions do represent a pseudo-Type IV orientation.

Many of planners' conflicts involve confrontations between this Type IV orientation and the Type III orientation of others who expect to see issues resolved through conventional politics. Planners' invocations of rationality, as the examples show, are efforts to define situations in ways that give them an advantage. Claims of "rationality" against "politics" represent an effort to establish Type IV, rather than Type III, rules for making decisions. While both types of power have the consequence of influencing others, Type III recognizes the self as an actor and source of power, whereas Type IV denies the validity of personal action and emphasizes the action of principle. Thus planners' arguments that "political" concerns are meaningless represent, above all, claims about legitimate sources of influence over decisions.[16]

The Type IV power of orientation appears entrepreneurial because of two claims many planners make. First, they assert that decisions should be based on the tenets of rationality and that planners are uniquely competent to assess the implications of rational norms in any particular case. Second, planners hold that their efforts should be measured, not by others' approval or support, but by their conformity to rational principles, of which planners themselves are the best judges.

Acting in this Type IV power orientation offers planners authority without anxiety. This action solves the riddle of Chapter 2: how power, particularly that of subordinates, may be "appropriately impersonal." It is possible to influence others—including administrative superiors, elected officials, and the public—in the name of abstract principles. Any conflict is not personal; it is simply a confrontation between these people and logical principles (as planners represent them). Further, the Type IV orientation offers a solution to the problem of performing without shame. Where there is no personal confrontation, there is no possibility of personal shame.[17] Moreover, insofar as both planners and their supervisors are measured by their conformity to rational principles, then the superiors may themselves be shamed for failure, and their subordinates may gain authority over them.[18]

The Type IV power orientation includes service to both abstract principles and group interests. When planners emphasize only the orientation toward abstract principles, they limit both the power and the collaboration that may be brought to problem solving. As noted, many planners with this orientation reject the power that can come from the expression and reconciliation of strongly felt social interests. Further, when these planners think of collaboration, they tend to conceive of a cooperative effort, presumably led by planners, to interpret norms of rationality in the context of specific plans or proposals. Although this collaboration may empower such abstract principles as rationality and the public interest, it still tends to exclude such social—political—aims as the maintenance of groups. Only when planners extend the Type IV power orientation to serving group interests does the orientation solve problems by enabling people to act collectively in ways not possible for individuals alone.[19]

When planners impose a narrow Type IV orientation on problem-solving efforts, not only is the potential collaborative power limited, but also the range of problems identified is restricted. A true Type IV orientation does not preclude a Type III orientation. Actors may seek to influence one another as advocates of different social interests and different abstract principles. For instance, some actors may emphasize economic efficiency and growth, in the service of business interests, whereas others may emphasize equitable distribution of economic product, in the service of the poor. Their efforts to influence one another permit a range of resources, including the organization of people and the mustering of arguments. Under such circumstances, in the absence of one party's overwhelming dominance of organizational resources, the plurality of frameworks offers the possibility of discovering a relatively broad range of problems, a considerable number of potential solutions, and a relatively inclusive coalition in powerful support of particular courses of action.

DEFENSIVE BUREAUCRATIC PROBLEM SOLVING

Technical expertise may contribute invaluably to analyzing and solving the complex social problems for which bureaucratic organizations are responsible. However, the psychological structure of bureaucracy encourages members to emphasize technical competence not only when it may assist with these problems, but also when it may help avoid anxiety. The defensive use of technical expertise interferes with formal organizational problem solving in two ways. First, the intensity of anxiety makes situations which evoke it seem more important than organizational problems. A worker will give primary attention to designing and implementing actions to avoid shame, guilt, or reprisal. The diffuseness of anxiety makes this domain seem to expand almost without limit. Claims of technical expertise may momentarily defend a practitioner from threats of anxiety, but the continual emergence of new threats requires constant organization of defenses. This process consumes considerable energy, time, and attention that would have been available for organizational problems.

Even when bureaucratic workers do consider these problems, the defensive focus on technical expertise creates a second difficulty. Solving a problem requires discovering a formulation of it which makes sense to interested parties and encourages them to give it further attention. However, when planners introduce technical expertise primarily as protection against anxiety, this emphasis, rather than increasing available information, resists new, potentially unsettling, information. In particular, it resists psychological, sociological and political frameworks in thinking about problems and their solution. Consequently, it becomes difficult to formulate problems workably, some efficacious courses of action may be overlooked, and some problematic situations may never become disentangled.

Schön (1982) offers a case illustration of this process with a planner who reviews development applications for a zoning board of appeals and recommends actions to the board. Board members assume that he relies on professional standards within the law; however, because the law and standards are ambiguous, he can use discretion to negotiate with developers. He wants to control development consistent with his beliefs about design, but he believes that he cannot openly acknowledge negotiating, because doing so would reveal that reviews are subject to personal judgment, for which he could be held culpable. Consequently, he negotiates without admitting this is what he does. He seeks to prevent either the board or developers from catching him exercising discretion by emphasizing his adherence to technical norms. Counterambiguously, he denies responsibility.

However, this position leads him to restrict information collection, prevents his finding a solution to a problem, and defeats a proposal which he himself favors. Because he is anxious about maintaining delicate control of concealed negotiations, he withholds and discourages any information which would identify problems not clearly governed by law and reveal his discretion to make recommendations. In this case, he tells a developer that a project requires a zoning variance. Although the planner regards the variance as a simple matter, the developer thinks it is a major obstacle. However, since the planner suppresses overtly political discussion, the developer feels that he cannot negotiate with him about his seemingly major problem, and he withdraws his application for the project, which, the planner also does not tell him, the planner himself favors.

Consciously, bureaucrats may believe that they openly confront problems and seek to collect a great deal of information about them. Plans for the defensive use of technical expertise are largely unconscious. Workers may not even be aware that they are simultaneously embarked on conflicting missions. Further, both the force and the covert character of defensive interests give them an advantage in competing with conscious problem-solving intentions.[20] The preceding chapter revealed that defensive interests lead workers to attempt to escape the bureaucracy in search of outside compensation. This chapter has shown that, even when workers do not actively escape formal responsibilities, concerns about defending their competence lead them into an unconscious realm in which those formal responsibilities are only secondary concerns.

NOTES

1. This discussion draws on the work of Argyris and Schön (1974, 1978; Argyris, 1982) in analyzing how practitioners solve problems. Argyris and Schön, studying more than 4,000 practitioners in numerous occupations, have found that, while one commonly espoused theory of problem solving emphasizes broad research and communication, the most common theory-in-use is closed and defensive. Argyris and Schön analyze their findings in a language of cognition and learning, but their work is consistent with psychodynamic formulations.

2. One might argue, however, that requirements of extensive, much less comprehensive, information collection and analysis are unrealistic under routine conditions of organizational decision making. See, for example, Downs (1967), Lindblom (1979), Simon (1976), Weick (1983), and Wildavsky (1979).

3. It is impossible to say precisely what proportion of planners emphasize technical norms to defend themselves against anxiety in the ways described here. However, empirical studies suggest that planners may be grouped roughly in the following way. A substantial minority, perhaps 20 percent, recognize political issues and actions as legitimate to planning and take explicitly political roles themselves. A large group, perhaps one-half, espouse straightforward technical roles: they collect technical information and analyze it without consideration of political interests. Another group, perhaps one-third, describe themselves inconsistently. While they recognize that political realities make it impossible for planners to act neutrally, they advocate technically neutral roles.

The last group certainly includes planners who are anxious about exercising power and defend themselves against this anxiety by choosing a technical role isolated from politics. The second group probably includes some like this as well, although they are less evidently ambivalent. They may be more successful at denying the inherent political meaning of their activities. Thus one might speculate conservatively that perhaps one-third to one-half of planners claim technical roles partly for the defensive reasons described here.

See Baum (1986) for a review of these studies of planners. There are no comparable studies of administrators or bureaucrats in other fields.

4. For good analytic summaries of the literature on professions, see Freidson (1970) and Larson (1977). For an examination of planning in these terms, see Baum (1983a).

5. A clash between hierarchical and professional authority is common in bureaucratic organizations. See, for example, Freidson (1970), Hall (1975, 1977), and Larson (1977).

6. Indeed, Schön (1983) shows that the creativity of practitioners in many fields rests on their ability to transpose frameworks, metaphors, and terms from their original domains onto apparently incongruous situations. This juxtaposition of concepts and problems stimulates insightful formulations of design solutions.

7. Indeed, generalism and autonomy may be inherently contradictory. Planners' effectiveness, based on general knowledge, requires the opposite of autonomy: consultation and accountability. For an elaboration of this argument, see Baum (1983a).

8. As noted earlier, an implicit model of problem solving is what Argyris and Schön have designated a "theory-in-use."

9. Lipsky (1980) portrays similar defenses among a wide range of "street-level bureaucrats." Although his analysis is not explicitly psychodynamic, he indicates that many bureaucratic workers, in addition to planners, seek to control—or ignore—clients in order to defend themselves against risks of punishment within the bureaucracy.

10. Interestingly, Cole (1975) studied planners' cognitive styles to determine the relative degree they think concretely or abstractly. Of those planners who could be definitely placed upon a concrete-abstract continuum, 55 percent were concrete thinkers and 45 percent were abstract thinkers. Thus planners' criticisms of the lay public may not accurately reflect planners' own cognitive styles.

11. Not all planners take such a position, but the implications of any assertion of a higher type of rationality is similar.

12. Altshuler (1965) reported that most planners claim to represent a "public interest" that transcends all partial interests. In a recent survey, Howe (1983) found that, although planners hold widely varying ideas about what the "public interest" is, most believe in it as a standard for their recommendations. In this view, although conventional interest group politics may illuminate components of the "public interest," planners reserve for themselves the prerogative of determining what represents the "public interest."

13. Not all planners regard political information as irrelevant to problem solving. Studies of them tend to find a minority who recognize either the propriety or inescapability of political matters (Baum, 1983b). However, any claims to expertise implicitly discount—or require interpretation of—others' statements. At issue here is not the value of expertise, which is certain, but the implications of claims of expertise for control of social meaning.

14. Private planning consultants have direct, financial relationships with their clients.

15. See Note 18 in Chapter 4.

16. Kohlberg (see Kohlberg & Gilligan, 1972; Kohlberg & Kramer, 1969) has studied decision making in terms of its developmental moral logic. The conflict described in this chapter as a conflict between a Type IV and a Type III power orientation would be considered by Kohlberg to be a conflict between two "postconventional" stages of moral reasoning: Stage 6 (a universal ethical principle orientation) and Stage 5 (a social contract legalistic orientation). For this complementary perspective on the issues posed here, the reader is referred to Kohlberg's work. For an illuminating analysis of the dynamics of an argument between groups in different stages of moral development, see Wiener (1973, chap. 5).

17. Merton (1940) was the first to suggest that the social structure of bureaucracy would lead bureaucrats ritualistically to invoke organizational rules and impersonal standards as a means of absolving themselves of responsibility for decision.

18. This emancipation from subordinacy is analogous to the personal meaning of the development of a Type IV power orientation during adolescence. This new orientation represents the transcendence of Oedipal conflicts with parents and movement toward relations more or less independent of them.

19. It should be clear that some planners and other bureaucrats, although a minority, do embrace the broader Type IV orientation.

20. For an analysis of the ways in which unconscious defenses elude discovery by the conscious mind, see Freud (1946). For a cognitively framed analysis of ways in which defenses may inhibit learning about problems which they create, see Argyris (1982).

7

The Advisor as Invited Intruder: A Case Example

As practiced by the Greeks of Asia Minor in the sixth century before our era, the custom of the scapegoat was as follows. When a city suffered from plague, famine, or other public calamity, an ugly or deformed person was chosen to take upon himself all the evils which afflicted the community. He was brought to a suitable place, where dried figs, a barley loaf, and cheese were put into his hand. These he ate. Then he was beaten seven times upon his genital organs with squills and branches of the wild fig and other wild trees, while the flutes played a particular tune. Afterwards he was burned on a pyre built of the wood of forest trees; and his ashes were cast into the sea. (James George Frazer, *The Golden Bough*)

Corporate nature abhors a vacuum. It also hates to pay benefits. Enter the consultant, who sells ideas superbly, totes good visuals, and may be fired at any time. Hard not to hire a guy like that and pay him a bundle in exchange for the transience of his position. Thus, while inside drones slave away over hot Wangs for reasonable wages, consultants float in the ether like angels, or vultures. Some carve careers out of simply taking meals, putting forth random notions over eggs and bacon, pastrami on rye, or steak au poivre with equal esprit. . . . When they are bad, they are very, very bad, and aggravating too. And out of control they are horrid. Total domination is the only way to play the game and win. These guys are loose cannons rolling around the corporation, and if they aren't tied down, they can go off in your face. . . .

In fact, judicious paranoia is entirely appropriate when a hired gun comes to town. A consultant always plugs a hole. Make sure it's not yours, or one you covet, because he just might get it if you don't watch out. . . . In short, ask not for whom the consultant toils, he toils for himself.

That's not to say he can't labor for you too. . . . Since he's unencumbered by sentiment or loyalty, he's comfortable scrutinizing the warts and arthritic joints of the corporation, the flaws no insider could stand to gaze upon for too long. . . .

Finally, consultants make consummate sacrificial lambs. When something bad goes down, firing a favored consultant can be just the act of atonement management needs to take you seriously again. (Stanley Bing, "The Strategist; Consultants and You," *Esquire*, October 1985)

A stranger is invited into the community to solve its problems. This outsider talks with members of the community, and he comes to know their hidden sins. The people feed him. And then they turn on him. With each blow they inflict on him, they believe they expel their sins. And then they feel better.

These two accounts, although referring to practices separated by millennia, are strikingly similar. They both describe a magical means of solving collective problems by sacrificing a symbolic repository of the problems.

Although people ask advice for many reasons, these stories suggest that one covert, perhaps unconscious, motive may be a wish to punish someone else for one's problems and anxieties. The next two chapters examine ways such a wish may enter into problem-solving efforts as an unconscious response to bureaucratic predicaments. This chapter analyzes a case example of how these punitive wishes may distort an advising relationship and divert it from its ostensible purpose of instrumental problem solving.

When bureaucratic workers face problems, they often ask others for advice. This consultation may involve formally designated, perhaps especially hired advisors or it may be an informal part of everyday interactions. A worker may voluntarily select an advisor or be told to get advice, perhaps from a specific person. When the advisor comes from elsewhere inside the organization, he or she may offer the benefits of familiarity with programs and procedures, whereas an outsider has the advantage of detached objectivity.

The focus of advice varies from specific substantive problems to conflicts in organizational processes. Sometimes a request which begins with difficulties in making a substantive decision turns into an investigation of organizational dynamics which interfere with decision making generally.[1] Some clients ask for specific answers to well-bounded questions; others prefer to work with consultants in trying to figure out what the problems are to begin with. Some clients want their advisors to stay in close touch; others expect their consultants to work relatively independently. Some clients expect advice to be narrowly delineated technical assistance, whereas others may look for broad guidance in organizational development.[2]

Despite all these variations, many advice-seeking relationships have two things in common, one concerning the relations between advisor and client, and the other concerning the problems on which they work. First, regardless of where an advisor comes from, in an important sense, he or she is an outsider. The problem "belongs" to the client, who has the responsibility for solving it; the consultant's only responsibility is giving advice. Second, whether someone asks for advice on a substantive problem or on organizational dynamics, many times the difficulties leading to a need for advice have to do with blocks in working relationships. Even advice focusing on substantive problems may become tacit process consultation.

These common characteristics of advice-seeking relationships affect problem solving in two ways. First, the division of responsibility, in which an outsider talks with someone about his or her problems, may give rise to further divisions which get in the way of thinking and talking about problems. Some divisions of labor are appropriate for solving problems. The confrontation between someone who has a problem and someone who looks at it is useful, because the former may have a strong investment in doing something, while the latter may have sufficient distance to analyze the options. Further, in any work group, task, emotional, and other leaders may become differentiated from followers (Bales, 1953; Bales & Slater, 1955).[3] However, people who have problems may unconsciously feel just enough hostility toward advisors who witness their difficulties that conscious investigative efforts and rational allocations of responsibility are subverted.

In addition, if the problem is embedded in organizational conflict, it is difficult to work on it without encountering the conflict; detaching the problem means ignoring it, not resolving it. If the worker with the problem feels hostile or anxious toward others, he or she will express these feelings somewhere, and the advising relationship provides an immediate, convenient, even safe place. The advisor who repeatedly touches nerves may seem like a reasonable target for feelings that started out in organizational battles.

The traditional literature on consulting recognizes that a client may consider organizational interests and relations foremost when requesting and using an advisor. However, this literature focuses on a client's conscious calculations, with little on unconscious expectations of and strategies toward the consultant.

PERSPECTIVES ON ADVISING

Most discussions of advising begin, at least implicitly, with an ideal image of consulting. Benveniste (1977), for example, says that "the primary function" of expertise is to give others information to help them resolve uncertainty about what to do. Meltsner (1976) notes that analysis provides information about the consequences of different choices.

An advisor represents a source of potential new intellectual and political power for someone who cannot manage a problem. Hence in the ideal consulting relationship a client does whatever is necessary and possible to empower the advisor to serve the client's interests and solve the client's problem. Central to this goal is candor; the advisor can help with the problem only if he or she gets an accurate picture of what it really is. The client provides the advisor with information about both the substance of the problem and the organizational and political context in which action may be taken. When the advisor asks questions, the client tells what he or she knows, and this information may begin to reveal ambiguities in the nature of the problem, the client's role, organizational decision making, or the advisor's role. Committed to finding a solution, the client even reveals his or her failures, shortcomings, and errors in judgment.

Studies of advising find this ideal violated more often than not. The most common explanations have focused on the political character of advice seeking. They emphasize how a client's political interests shape communication with the advisor (see Archibald, 1970; Benveniste, 1977; Meltsner, 1976). In this view, the bureaucratic client is a rational actor, calculating political interests and assigning the advisor tasks which serve them. Further, because the client is political, he or she conceives the relationship with the advisor similarly, weighing whether the advisor may be an ally or adversary.[4]

In this perspective, a bureaucratic client may withhold information from a consultant for two self-interested reasons. First, the client may be concerned about a problem which he or she feels the consultant might not approve of or could not be trusted to keep secret. For example, the client's intentions may involve personal ambition, treachery, a publicly unpopular agenda, ethically questionable aims, or strategically sensitive actions. Accordingly, the client may freely describe an ostensible problem to the consultant, interspersing just enough about the "real" problem to get useful advice on it without revealing it. In turn, the client may plan to use only a small part of the information the consultant offers.

Second, the client may withhold information from the advisor because of covert intentions about the advisor, what Benveniste (1977) calls "the secondary functions" of expertise. The client may want to use the consultation as a way of stalling until, perhaps, a problem goes away. Hence the client may not really take the consultation seriously. Alternatively, the client may want to use the advisor's work symbolically, to legitimate what the client wants to do all along. Thus the client may have a "correct answer" in mind and feed the advisor only information leading to that answer, or the client may simply use the fact of a consultant's study, regardless of its content, to legitimate the final choice of action. Finally, the client may want to use the consultant to present new ideas, with the advisor running any risks and taking any blame.

Even in cases where a bureaucratic client seeks advice relatively openly, an advisor may present problems that make it difficult to discuss situations candidly. For one thing, an advisor may also have political interests and may promote what Benveniste (1977) has called "the politics of expertise." The advisor, desirous of getting the client to take certain courses of action, may couch advice in ways designed to favor those actions—even when they might not be in the client's interests. Thus a client may fear that the advisor has not investigated all possibilities, has withheld information, or has covertly built a coalition in favor of certain proposals. In response, the client may become more secretive or strategically selective in passing on information.

Alternatively, the advisor may be loyal, but may produce incomprehensible or useless information. The consultant may not understand the expectation that advice serve the client's political interests. One common formulation of this difficulty emphasizes that people who become advisors tend to have different cognitive styles than political actors (see Benne, 1976; Churchman & Schainblatt, 1965; Doktor & Hamilton, 1973; Huysmans, 1970; Mintzberg, 1976). Whereas bureaucratic clients usually think in terms of finding practical solutions to specific short-range problems, their advisors—particularly those with natural or social scientific training—may be accustomed to looking for broad context-free theoretical generalizations encompassing a range of cases. Thus consultants may offer suggestions that are either highly abstract or jargon-laden, and in neither case helpful in solving a tangible political problem.

All these possibilities reduce a client's willingness to speak candidly with or empower the advisor. And yet they represent only the surface complexities of advice-seeking relationships. A psychoanalytic perspective goes deeper, examining clients' unconscious interests in this relationship. In particular, the psychoanalytic view looks at ways in which these unconscious interests may interfere with and defeat the strategic calculations emphasized in the political perspective.[5] The two basic characteristics of advice-seeking relationships identified earlier—tensions about responsibility between client and consultant, and diversion of organizational stresses into consultation—both arouse unconscious interests and wishes in relation to advisors. These interests lead to communication distortions and diversions from conscious problem-solving.[6] Additionally, these wishes lead the client away from empowering the advisor toward incapacitating the advisor.[7] An advisor may respond to such mixed intensions by seeking simply to keep the client under control or by giving up competence to help.

The following case example illustrates the mixed feelings with which clients seek advice. Many actions characterized here as unconsciously motivated may also be interpreted as consciously political. Two points should be made about the latter interpretation. The first is that ostensibly "rational" notions of conscious political action often tacitly incorporate unconscious assumptions about organizational relationships and are a response to bureaucratic psychological structure. Second, it must be acknowledged that efforts to weaken advisors may have more than one motivation.[8]

A CONSULTATION

Mr. Simon is a private consultant who advises public agencies on planning issues in health care and social services delivery. Although requests for his assistance normally focus on substantive problems, he believes that organizational relationships, including unconscious assumptions about them, influence and constrain the possibility of dealing with substantive problems he is asked to solve. Accordingly, he watches for clues about unconscious organizational life and tries to respond to covert issues while discussing overt problems. When he does call his clients' attention to some covert issues, he avoids psychological language.

The following case involves a request for advice on reorganizing a service program. The consultation rests on the boundary between a substantive planning consultation and an organizational process consultation. In many respects the consultant acted effectively; unexpected events at the completion of his service leave unanswerable questions about what results he might have had under other circumstances. Still, the consultation is not presented as a model. The account here focuses on the process of the consultation, rather than its outcome, and it is clear that the consultant makes some questionable decisions. The consultant's inclinations and actions throw light onto the dynamics of advice-seeking relationships under the common conditions described earlier. I present this case because I have had extended access to the consultant and facts about his work.

The consultation involved an interdepartmental health services planning program in a highly urbanized county. The county executive's human services director had requested that heads of all departments with any connection to health send representatives to 25 small area health planning councils, so that the county might coordinate services in a context of fiscal austerity. The director had no staff member responsible for overseeing the program, and he directed it only sporadically when he had time or when he needed to use it to distribute patronage for the county executive. Councils had met for approximately two-and-a-half years with desultory results when he asked for advice.

The director needed conspicuous results from the councils for two reasons. First, he had started the program, and had now involved almost 300 county workers in biweekly council meetings, without any significant outcomes. In addition, as the county executive's human resources director and a successful black politician, at a time when the executive was being criticized for favoring middle-income voters against the poor, the director needed the program to produce noticeable results for the poor, among others. He asked Simon to ascertain what the councils were doing and to advise him on how the program could be improved.

Simon readily discovered several reasons for the limited activity of council members, most of which centered on an asymmetrical distribution of responsibility and authority and on the director's ambiguous, autocratic management style. The director had no formal authority over the city's operating departments but told department heads to assign staff members to 25 councils. Department heads attached low importance to the councils and generally delegated low-status staff members to them. Council members then found themselves confronting ambitious assignments from the director to reform the health care system, while they had no real authority to carry out these assignments. Because the director lacked formal authority over council members, he tended to substitute intimidation for a tangible system of incentives. Thus council members were further discouraged from taking initiatives by the director's style, in which he alternately exhorted them to be creative, delivered emergency requests to drop everything and help out with a special project from the county executive's office, and encouraged timely, conspicuous innovations. More could be said about these issues. They are certainly not unique to this program or setting. However, the analysis will focus on the process through which the executive's office requested advice. This account focuses on three puzzling incidents.

Incident One: The Seduction

In presenting the consulting assignment, the human services director suggested that Simon work on a day-to-day basis with the assistant director, a black woman with ties to the black medical community. One of the first times Simon met with her, she remarked that he looked familiar, that she must know him from somewhere. After reflection, she recalled that he reminded her of Jose Ferrer. Simon was flattered, although he had never been likened to Ferrer before and had never even seen his movies. She mentioned this resemblance again on another occasion, when she said that he reminded her specifically of Ferrer in his role as Cyrano de Bergerac. Simon wondered whether the mention of Cyrano were not at least an unconscious reference to his nose, as a symbol of his being identified as Jewish. Nevertheless, he appreciated both the unexpected compliment and the attempt at familiarity.

In the first weeks of work Simon interviewed chairpersons of some of the councils. One prominent theme in their comments was apprehension about the ambiguous, apparently arbitrary authority of the director. When the assistant director asked him what the chairpersons were telling him, he summarized these themes. She agreed that the director's management style was an issue the consultation should address, and she added her own complaints about him. When she met Simon after he had done more interviews, and he reported similar statements from still other chairpersons, she warmly agreed about the validity of what he was hearing and observed, "Now you are one of us."

The assistant director's actions expressed a combination of conscious and unconscious intentions. Her friendliness was conscious and probably genuine, as was her interest in developing a cooperative working relationship. However, her actions apparently also expressed unconscious intentions that shed some light on organizational dynamics, and can serve as an example of relationships that may commonly develop between client and advisor. The central aspect of the drama initiated by the reference to Jose Ferrer is seduction.

The comparison of the advisor to a familiar movie star is extremely flattering; particularly coming from a relative stranger, it invites a reciprocal response—if not a similar compliment, then a favor of some kind. The obvious favor would be sympathy with the client's point of view.

At the same time, it is clear that the assistant director had a network of organizational relationships. For example, she was sympathetic with council chairpersons who were hostile toward her boss for being autocratic. She asked for her favor in this context when she observed that the director's management style should be addressed by the consultation. She wanted the advisor to take her side in her conflict with her boss. Here it should be recalled that her boss was really the formal client. In her effort to enlist the support of the advisor, she attempted to make the formal consulting relationship sufficiently ambiguous as to permit her to act as the client. She attempted to cement this new alliance by telling the advisor that his newly acquired understanding now made him "one of us": the same in thinking—and, presumably, interests—as the assistant director and other organizational subordinates.

Still, the indirect reference to the advisor's nose pointed to a persistent difference between client and advisor. It was not necessarily an allusion to the advisor's Jewishness, although the fictional Cyrano expressed his love for Roxane through the more handsome Christian. At the least, Cyrano's nose was an ugly marking which prevented him from openly wooing Roxane. Instead, he became Christian's advisor, giving him the words with which Christian successfully courted Roxane. Only as Cyrano was breathing his dying words to Roxane, did she recognize that he, not Christian, had known her most clearly (Rostand, 1890; rep. 1951). The fleeting reference to Cyrano de Bergerac, then, called attention to a significant difference between the advisor and the client and, although intimating that the advisor might understand the client deeply, portended little ultimate success for him.

The next incident represents an effort by the assistant director to make use of the advisor in her proposed alliance with him.

Incident Two: The Setup

The next time Simon and the assistant director met, she suggested that he report his findings the following week at the monthly meeting of the 25 council chairpersons. About this time, the director had appointed a full-time coordinator for the program, a black man with long ties to the county's antipoverty program, and he was present at this discussion between Simon and the assistant director. He also encouraged Simon to report to the chairpersons. The assistant director went on to observe that, because the director's presence at the monthly meeting intimidated those attending and stifled discussion, she would take the extraordinary step of asking him to refrain from attending, so that Simon and the chairpersons might candidly discuss the program.

Simon was apprehensive about a premature presentation of his data, not simply because his tentative interpretations might be inaccurate, but also because those interpretations emphasized the director's responsibility for the program's difficulties. Nevertheless, the assistant director pushed him to appear, and Simon gave in, deciding to present his report as simply a summary of what some council members had told him.

The director, indeed, did not attend the meeting. Simon presented his findings, taking care to explain how the program's incentive structure (rather than the director personally) had contributed to some difficulties. Slowly, some of the chairpersons began to comment, generally favorably. One black man observed that he was pleased to find that this report, unlike that of most consultants, was not "a whitewash." Simon took this racial double entendre to be a sign of approval. People began to raise probing questions about his analysis of the program, pushing beyond the areas covered in his presentation. As he was about to respond to one particularly complicated question, the door of the room opened behind him, and the room became instantly still. The director had come to the meeting after all. At this moment, the chairpersons had the probable pleasure of hearing Simon indirectly express some of their criticisms to the director of the program. After Simon's response, the chairpersons said little else, and the director concluded the meeting by reminding them that the program was extremely important.

After the assistant director's attempt to lure the advisor over to her side against the director, her request to the director not to attend the chairpersons' meeting could be interpreted as a growing conspiracy. However, in the meeting the consultant ended up reciting a number of her charges to the face of the director. The intervening events were that she had invited the advisor to make a presentation incorporating criticisms of the director and she had told the advisor that she would ask the director not to attend the meeting. She had accompanied her invitation with the assurance that the advisor's observations were certainly valid. She probably did ask the director not to attend the meeting; however, she might have anticipated that he would find it difficult to stay away from a meeting in which his consultant would present preliminary findings.

If all the assistant director's actions had been conscious and deliberate, one might conlude that she wanted to put the advisor into a position in which he not only criticized her boss publicly, but did so to his face. If she, indeed, believed that her boss was punitive—and the general reaction to the director's entrance certainly confirms this view—then one might interpret her actions as putting the advisor into a position where he would criticize her boss and then be punished by him.

The first two incidents present a curious sequence. In the first incident, a member of the client organization attempted to seduce the advisor to take her side against her organizational superior. In the second incident, the subordinate set up the advisor to present her case against her boss, even though taking her side apparently meant the sacrifice of the advisor, who could no longer be useful to anyone in the client organization. A third incident offers a clue to this puzzle.

Incident Three: The Death Threat

A few days after the meeting with the chairpersons, Simon phoned the assistant director to review his work. She brought up the meeting and observed that the director did not "understand the issue of his power"—he intimidated others and insensitively resisted constructive ideas that did not match his current political concerns.

Then, abruptly seeming to change the subject, she mentioned that two years earlier the director had engaged another consultant for a similar assignment. Simon was amazed that, until this moment, approximately two months after he had accepted the consultation, no one had mentioned this other advisor. She went on to say that, after five weeks of work, that consultant had written the director a letter questioning the premises of the program. From her description of the letter, Simon thought that the criticisms and recommendations the earlier advisor had made were similar to those which he had made or might make. In response to the letter, the director "was furious" and fired the consultant. Since that time, he had never been mentioned again, and his letter had simply been filed in the director's office.

Simon felt angry that no one had ever mentioned this consultant or the letter to him before, and, because what the consultant had said resembled his own thoughts, he was anxious to understand why the other had been fired. He anticipated that the assistant director would offer him a copy of the letter, and while she talked, he pondered the pros and cons of accepting. To accept would be clear collusion with the assistant director, and he did not want to incur any obligations or problems along those lines. However, he also believed that, if he wanted to have any impact on the program, he needed to know what was in the letter and what had been objectionable in it. He chose to believe that if he accepted the letter, the director would never know. Soon the assistant director indeed offered him a photocopy of the letter. Simon accepted the offer, and she asked him for his home address. She then gave him the name and address of the earlier consultant.

When he phoned the other consultant, they talked amiably and shared observations about the program. Their analyses of the program were similar. The earlier consultant suggested that his undoing had been partly a result of his analysis and partly a result of his bluntness in pointing out erroneous assumptions. When Simon received the copy of the letter in the mail, he found little exceptional about it. The letter was perceptive and candid, but also encouraging. Its only possible fault, perhaps, was identifying problems to a man who preferred to know of none. Simon was more intrigued by the assistant director's note accompanying the letter. It said, "You've never seen, read, or discussed this. In fact, you don't even know me. Hope this info helps to understand the process of the councils." The note fascinated Simon with its hints of great plotting. He was flattered to be included in the assistant director's apparent confidence but also puzzled about the reasons for concealment of information useful to his assignment. It was as if concealment of the earlier consulting relationship were more important than the success of the present one. Simon ignored his apprehension at what the consequences for him could be, either for participating in this apparent plot or simply for being a consultant.

This communication from the assistant director to the advisor had some connection with the chairpersons meeting, because they were close in time and were discussed in the same phone conversation. The meeting had concluded with the advisor in a situation where punishment could be imagined. That episode could be regarded as at least a symbolic mortification. However, the third incident presents a possibility of yet more dire punishment: the advisor could be fired.

This last incident reveals two conflicting intentions toward the consultant.

Overtly, consciously, the assistant director acted to empower him in solving a problem (albeit her way) by giving him vital information about how to approach her boss, the actual client. The note's message of "look what (eventually) happens to people like you" was apparently a friendly warning, in which she wanted to alert the consultant to potential hazards in working for the director. She took real risks in passing on the earlier letter. If the consultant ever talked about it, she could deny her involvement, but she would have difficulty convincing the director that she had not leaked it. Thus, in disclosing the earlier consultation, she expressed political trust in the present consultant, including conspiratorial instructions to keep quiet about this matter.

At the same time, however, the note conveyed conflicting, perhaps unconscious, messages. After all, it was the assistant director who had set the advisor up for a confrontation with her punitive boss in the preceding incident. Further, in her note she implied but did not guarantee confidentiality on her side. In addition, crucially, she threatened that if the advisor ever broke the vow of secrecy, she would abandon him. The first line of the note was apparently a conspiratorial direction; the second line, however, conveyed a veiled threat: should the consultant's possession of the letter ever become public, the consultant would no longer be able to say that he knew the client. More directly, the client would no longer acknowledge knowing the advisor.

With this warning, the assistant director uttered two types of death threats and indicated her ultimate loyalty. When she first met the advisor, she had insisted to him that she "knew" him from somewhere, that they were already in some sense intimate. Here she threatened to withdraw that knowledge, the intimacy, and the implied alliance. On top of this, the threat no longer to know the advisor pointedly designated the fate of the earlier consultant, who was fired and now forgotten. Both threats made clear that the assistant director, while working with the advisor, even in conflict with her boss, ultimately regarded her primary loyalty to the boss and the organization, not to the advisor. This information, as the note's last line suggested, should help the consultant understand the blunt reality of the process of the councils.

Interpretation of these three incidents can illuminate certain aspects of organizational dynamics and the use of advisors. For one thing, members of a client organization may attempt to draw an advisor into alliances against their adversaries. In addition, members of the client organization are likely to feel stronger loyalty to their organization and colleagues, with whom they have long (even if ambivalent) relationships, than to advisors, whom they hardly know and who have little tangible to offer them. However, Simon's story still poses two puzzling questions. First, why would a client go through the effort of seducing an advisor into an alliance and then deliver the advisor up to punishment, perhaps even by the client's adversaries? Second, why would a client sever a relationship with an advisor (as in the case of the first consultant) when doing so would prevent solving the problem for which the client requested help?

Two aspects of these incidents offer clues. First, the repetition of the theme of the sacrificed advisor (the earlier advisor and then Simon) suggests something ritualistic about these actions. That is, there may be some persistent organizational problems for which members believe that periodic sacrifice of an advisor provides a remedy. Second, the counterpoint of conscious and unconscious intentions in these incidents, especially in the simultaneous efforts to empower and to expel the advisor, suggests that members' unconscious feelings and wishes about the organization may create some of the problems and call for this ritualistic cure. The exploration of these possibilities begins by examining the conclusion of Simon's consultation.

THE CONCLUSION OF THE CONSULTATION

The consultation was apparently successful. His presentation to the director and written report openly described problems in the unequal distribution of authority and responsibility, which discouraged anyone from taking responsibility. His recommendations emphasized redistributions of authority that could benefit both council members and the director. Simon said little about the director's management style, which he believed was unlikely to change.

The director praised the report and said that he wanted Simon to present a copy to the county executive at a special breakfast meeting. Before the meeting with the executive could be arranged, however, Simon read in the newspaper of the director's imminent departure from county government to another organization. He was succeeded by a man who had much less political acumen or ability, who had little familiarity with the council program, who could see little evident use for it, and who initially concentrated simply on learning his job.

The program continued under the direction of the recently appointed coordinator, who repeatedly attempted to persuade the new director of its importance. Simon's report was referred to rather widely in the small world of the county executive's office and the councils. Council members who saw it regarded it as fair and insightful. The executive's staff referred to "Simon's Report" with almost biblical connotations. The report did seem to offer genuinely effective ways to remedy the moribund council program. In the end, however, few of the approximately 40 recommendations, including some with no political or economic costs, were implemented. The coordinator finally succeeded in getting the new director's support for experimental modifications in the program. However, although the coordinator invoked the report as the authority for his changes, his actions violated all of its basic principles.

BUREAUCRATIC RESPONSIBILITY AND
EXPECTATIONS OF ADVISING

The bureaucratic setting influences a worker's relationship with an advisor by creating role expectations for the worker. In particular, the predicament of ambiguous and unequal responsibility and authority makes it difficult to ask for advice and risky to accept it. As noted, problems are often ambiguous. Consequently, when a worker acquires responsibility for a problem, he or she could benefit from help simply in defining the problem. However, its very ambiguity makes it difficult to ask clearly for advice or evaluate the information offered. Although a bureaucrat may accept the discomfort of not having the solution for a problem, he or she may feel embarrassed at not even knowing quite what the problem is. Moreover, as studies reviewed earlier (see Cohen, 1959; Kahn et al., 1964; Weiss, 1980) suggest, workers in ambiguous situations tend to have low self-esteem and doubt that they know what they are doing. For these reasons, a bureaucrat may avoid asking for advice, and when the worker does, he or she may feel ashamed about having a problem and anxious about appearing inept.

However, neither organizational ambiguity nor attitudes toward advisors are so simple. As Chapter 3 revealed, when both responsibility and authority are ambiguous, responsibility appears to grow ambiguously large, while authority becomes ambiguously small. Not only are workers uncertain where the bounds to their responsibilities lie, but they also tend to believe that power over decisions rests somewhere outside their control. The autonomy of bureaucratic authority complicates this situation both by confusing subordinates about the nature of organizational power and by disabling them from striving for it.

In this situation, exercising responsibility may only contribute to troubles. One may do a great deal of work and yet find that others make the final decisions. If the decisions solve problems, presumably others will receive the credit. However, bureaucrats' worst fears concern the risk that decisions lead to errors. If a worker bears responsibility for tasks while others have ultimate power, then it is always possible for others to hold the worker culpable for failure. In this way, the ambiguities of a bureaucratic role contribute to the fear of being scapegoated for problems.

The ideal solution is to have decision-making power while being able to disavow responsibility when failure results. For example, the last chapter described workers claiming technical expertise in order to create counterambiguity to undermine others' understandings about problems. Ritualism enables workers to acquire power but transfers responsibility for decisions to external rules.

Still, the difficulty with such efforts, beyond costs to the self-esteem of workers wanting recognition, is that subordinates can rarely evade all responsibility imposed by superiors. At best, a subordinate may attempt to see that blame gets settled on someone else. One good candidate for such scapegoating is someone lower in the hierarchy who already bears some responsibility and, by definition, wields less power than the worker concerned. Alternatively, another good candidate is an advisor, who exercises no formal power, may be directly accountable to the bureaucratic client, and might be persuaded to assume responsibility for decisions as a compensation for a lack of formal power.

Workers' fears of being scapegoated lead to mixed impulses toward an advisor. On the one hand, advice which helps to define, delimit, and carry out responsibilities is desirable, and if it promises success, the client may attempt to take the credit. On the other hand, insofar as the worker lacks authority to match the responsibilities, then an advisor's work may be useless or even dangerous—useless if the bureaucrat's work does not directly affect final decisions and dangerous if it makes the worker prominent and more likely to be blamed if subsequent decisions lead to failure. These latter possibilities discourage seeking advice. However, if the client must act, then he or she may attempt to tie responsibility for implementation to the advisor. In this way, the advisor may appear to be the cause of failure.

An advisor may facilitate such a transfer of culpability—in effect, becoming the scapegoat for organizational problems—by wanting to take responsibility. Someone who has invested energy and time in preparing recommendations would like the satisfaction of seeing them implemented, of getting a sense of completion to work, and of receiving credit for what he or she assumes will be success. Thus the advisor may want to take responsibility at the same time the client may want to shift it away.

TENSIONS WITHIN THE ADVISING RELATIONSHIP

This advice-seeking ambivalence aroused by the bureaucratic psychological structure is reinforced by typical tensions in advising relationships, which lead clients to disable their consultants at the same time that they seek to empower them. The primary sources of tension are a client's shame about having problems, perception of the advisor as an outsider, and dependency on the advisor.

The Shame of Problems

As already noted, many people feel ashamed about having problems they cannot solve alone. They regard requesting advice as enlarging the audience to their failures and rendering themselves vulnerable to still someone else's evaluation (Forester, 1981). This possibility of judgment adds anxiety (Dickey & Hampton, 1981), heightens any feelings of self-doubt, and encourages anew a repertoire of defensive responses to potentially shaming criticism.

The Advisor as Outsider

An advisor is normally an outsider whom one admits into one's work. Even if the consultant is a colleague, he or she is usually from another part of the organization and probably knew little or nothing about the client's work until called in. Furthermore, the client retains responsibility for making and implementing decisions. The advisor and client thus have different past knowledge, present responsibilities, and future risks.

Work with an advisor begins with a revelation of how one has failed to carry out one's responsibilities, or, at least, how one cannot fulfill them alone. Even with an attempt to cover any personal blame or ineptitude, the account must be sufficiently plausible to persuade the advisor to accept it. A consultant may anticipate receiving an account that deliberately or accidentally conceals embarrassing facts, hidden agendas, or unconscious ideas. The advisor may suspect that the real problem for which a client wants help differs from the problem presented and may believe that success depends on breaking down the client's presentation. The client may anticipate this skepticism from past relationships with advisors, adding to the ambivalence of the relationship. In short, a client cannot automatically trust an advisor to accept a problem as defined.

Thus, whereas one asset of an outsider is the detachment that enables the person to see what insiders cannot, a client may fear that an advisor will indeed see—and talk about—some things the client would prefer to keep invisible. An outsider may not take for granted certain everyday organizational assumptions (for example, that a manifestly inept worker is competent; that two departments seeking to destroy one another are collaborating; that subordinates' weak performances are the best they can do and not a passively hostile response to what they regard as their boss' intimidation; or that certain problems are inevitable). Although some of these assumptions may interfere with carrying out solving problems, they may more or less successfully protect members from anxiety that would arise in confrontive personal contact. Thus a client may prefer to preserve the assumptions and their benefits rather than undergo the embarrassment, anxiety, or more tangible loss of having their erroneous foundation revealed. These assumptions may be accepted "secrets" within the organization—matters taken for granted and not open to discussion, no matter what their relevance to organizational work. An outsider cannot be depended on to recognize that these "secrets" are exempted from examination, or at least to agree to respect their status as "secrets."[9]

The advisor, on the other hand, may believe that useful consulting requires just this unsettling of organizational assumptions. So, while offering the potential for solving a client's problem, the advisor threatens to disturb organizational balances. Even though the advisor is an invited outsider, advising continually intrudes into the client's work and assumptions. Thus the invited intruder offers help but also poses a danger.

Dependency on the Advisor and Other Assumptions

Organizational responsibilities, anxiety about being shamed for problems, and vulnerability to outside scrutiny make a client intensely dependent on an advisor. The client depends on the advisor's help to solve a problem, the advisor's judgment for vindication from any culpability, and the advisor's discretion to respect the client's assumptions and dignity.

Accordingly, a client would like to know enough about a consultant to feel that he or she can be trusted. However, insofar as the advisor is an outsider, a client must usually resort to imagination, at least initially, to supply information about this person on whom so much depends. Thus a bureaucrat may engage in transference with an advisor similar to the process with organizational superiors. Dependency on an advisor may unconsciously evoke memories of early parental relationships and may introduce fantasies into the advice-seeking endeavor. Some of these fantasies may work unconsciously against conscious intentions to collaborate with a consultant.

The Advisor's Paternity

Unconscious transference of childlike feelings to advisors is an ancient phenomenon. One of the first recorded examples involves the biblical account of the Egyptian pharaoh's appointment of Joseph as his advisor. Joseph was not a member of the royal family. After interpreting the pharaoh's dreams as portending a famine in Egypt, Joseph was released from prison by the ruler, who then made Joseph his chief advisor, overseeing, among other things, preparation of granaries to tide Egypt through the lean years. When Joseph reflected on his new status, he designated the position as "Ab"—literally, "father"—to the pharaoh (Genesis 45:8). Thus at least the advisor regarded himself as father to the client. Indication that the perception was reciprocal comes from evidence that "Ab" was an exact Hebrew transliteration of an Egyptian title of state rank corresponding to "vizier" (Hertz, 1958).[10]

A client is particularly likely to think of an advisor paternally when the advisor is a man, although a client may also think of a male in some maternal terms.[11] As with fantasies about administrators, transferences to consultants may include both positive and negative assumptions. Positively, a consultant may seem strong and dependable, perhaps also nurturing. Negatively, a consultant may seem oppressively powerful. In addition, when an advisor seems parentlike, the client may unconsciously worry that the consultant is more loyal to the fantasied other parent rather than to the client. An alternative positive transference may involve the fantasy of no longer being the parental advisor's child, but becoming his or her partner, as if to have together an imaginary child. Although common, these unconscious transferences are unrealistic, and each interferes with collaborative problem solving in different ways.

Exaggeration of Dependency

Positive parental transference to an advisor often initially leads people who request advice to feel less competent than they really are, act helpless, and exaggerate their dependence on the advisor, to whom they attribute overwhelming ability. In short, they believe only the advisor can solve their problems. Bion (1961), observing a similar phenomenon repeatedly in work groups, called it the "dependency assumption." He observed that when group members become anxious that work on problems could result in unsettling changes, they unconsciously seek to defend themselves from work by making imaginary assumptions about the group and its leader to justify their inaction. One typical assumption is that the group members are extremely dependent on the leader, who alone is considered competent to act. People make this kind of assumption about the power of leadership and the impotence of followers in many types of groups, including the small groups created when one or more clients engage an advisor as a problem-solving leader. This assumption is an example of a more general tendency to avoid examining threatening problems, a phenomenon psychoanalysts call "resistance."[12]

This transference contributes to mixed feelings toward an advisor. Childlike affection for a strong parent reinforces the realistic belief that there are practical reasons for engaging the consultant's assistance. On the other hand, the security this dependency offers goes against organizational and professional norms, as well as dominant images of adulthood, which encourage independence and self-sufficiency.[13] Thus a client may also feel humiliated and resent the advisor, and may become angry toward the parentlike consultant as well as anxious about retribution, just as with organizational superiors. Further, because unqualified faith in an advisor's ability is unrealistic, disillusionment is likely, and a client may then grow still angrier at the advisor. In order to feel self-assured and secure, a worker may resist any consultation, hope the advisor fails, and attempt to frustrate any assistance from the advisor.[14]

Protecting the Maternal Organization from the Paternal Intruder

The first stage in instrumental work by any group entails defining the group's membership and boundaries (see Tuckman, 1965). Thus if a bureaucratic client were to move beyond the thicket of transferred dependency to attempt instrumental work, he or she would immediately confront the advisor's ambiguous position in relation to the organization.

At the beginning of a consultation a client has little overt evidence about an advisor's loyalty, and it may be reasonable to move tentatively and observe the advisor's responses. However, a bureaucrat feeling anxious about being blamed for problems may want to come quickly to a definite conclusion about the advisor. In such a case, a worker may turn to fantasy to supply additional "information." A common unconscious means for establishing boundaries, already discussed, is denying and splitting off one's undesirable characteristics, attributing them to someone else, and then seeing this bad person as separate from one's good self. A bureaucrat may already have drawn such distinctions between him- or herself and organizational authority. The worker may simply draw the consultant in as part of the "good" group against the "bad" administrator. Realistic needs for advice, plus any positive information about an advisor's past accomplishments, would support such an unconscious decision. If a worker unconsciously draws images of the advisor from the worker's picture of administrations, this transference would make the advisor seem like the bureaucratic boss and create doubt about the consultant's loyalty. If the worker opposes and fears a bureaucratic administrator, the worker may assume that the advisor is also to be opposed and feared, and the worker may put them both in a separate group.

At the same time, two basic realities—that the organization pays the worker and that the advisor is probably only temporary—may encourage the conclusion that the consultant is an outsider and not dependable. At the least the consultant is unlikely to share the client's long-term risks. Unconsciously, a client may reinforce these facts with new assumptions, now negative, about the advisor's parentlike qualities.

For example, when group members ostracize a leader, they often rationalize their actions with unconscious feelings that the leader is a cruel, depriving father who intrudes upon the group, which, save for his presence, would be supportive, loving, and undemanding (see Bion, 1961; Chasseguet-Smirgel, 1985; Slater, 1966). Essentially, group members expel their formal leader from their moral community. In similar ways, a bureaucratic worker may tacitly think of the organization, which pays him, however it really operates, as a nurturing mother, while the advisor seems like a distant and ungiving father, who perhaps disciplines, but is certainly not supportive and is possibly unfaithful. Thus a client may unconsciously resolve ambiguities in the advisor's position by thinking of the advisor as an intrusive assailant against a maternal organization.

Although it is reasonable to deliberate about who is in and outside a group, the advisor's position and transferences it evokes may lead the client to perseverate about boundary issues, rather than treating them as ancillary to other tasks. Such efforts to draw boundaries by identifying external enemies are examples of Bion's "fight-flight assumption" in groups, where members believe that fighting or fleeing an external enemy is the same as, or more important than, doing instrumental work. Vacillation about where to place the advisor in relation to organizational authority reflects the complexity of unconscious "information" about leaders. In the end, an organization's financial "nurturance" of its employees may most strongly support the assumption that the organization is motherlike and that the consultant is an assaultive intruder. These fantasy thoughts are all examples of resistance, in which clients avoid risks of instrumental work with consultants.

The Expectant Couple

Understandably, a client may prefer to find an advising relationship unambiguously collaborative. On the conscious level, norms for the ideal advising relationship encourage such optimism. These wishes may be bolstered unconsciously by new fantasies about the consultant. Rather than seeing oneself as a child to a parental advisor, a client may envision him- or herself as the consultant's partner. This thought may stem from early childhood wishes to have unquestioned affection and loyalty from the parent of the opposite sex. This assumption provides security for the worker, who sees the advisor as an equal, perhaps almost undifferentiated, partner and expects the relationship to be productive.

For example, group members often euphoriously expect magical solutions will be born from the imagined coupling of members, in what Bion called the "pairing assumption" (Bion, 1961; Slater, 1966). In these situations, group members do not recognize anyone among them as a leader but unconsciously assume that the imaginary coupling will produce a messianic future leader. All that group members need do is interact appropriately, and this leader will come to take responsibility for solving problems.

Similarly when confronting an advisor, a client may tacitly assume that their association will somehow culminate in creation of leadership and a solution. However, this utopian view evades work risks by avoiding a realistic assessment of the resources and potentials of consultation. Indeed, continuing optimism is possible only because the client never puts expectations to an empirical test. In time, a client inevitably finds this assumption disappointing and may turn in anger against the consultant.

The Sum of Expectations

These unconscious concerns and assumptions coexist with conscious intentions to work instrumentally, entering advice-seeking relationships in varying combinations and sequences. For example, a bureaucratic worker may begin by wanting a consultant's help in solving an important problem, quickly come to feel extremely dependent on the consultant, switch to seeing the consultant as a hostile intruder, turn to meaningful data collection and analysis, and then assume a placid pattern of neutrally optimistic interaction with the consultant. Which beliefs dominate a client's actions depend in part on the client's personality, but also on the composition and dynamics of the group involved, including members' personalities, styles, interactions, organizational relationships, and needs and expectations in relation to the problem.[15]

Bureaucratic workers are particularly likely to make these unconscious assumptions because each provides an imaginary defense against the shame of having problems and anxiety about being blamed for mistakes by redefining the advising relationship. When the unconscious assumptions dominate actions, they work against conscious problem-solving intentions in three ways. First, in redefining the consulting relationship, they point toward magical, rather than instrumental, solutions for problems. For example, total reliance on the advisor, hostility toward the advisor, or wishful interaction with the advisor may seem more important than brainstorming or data analysis.

In addition, these assumptions, while reformulating problem solving so as to be less anxiety-laden, may nevertheless redefine the advisor in a way that evokes new anxiety. For example, dependence on a consultant may make a client as resentful and anxious as dependence on a hierarchical superior. While thinking of an advisor as an intruder distracts a client from anxiety about real problems, anticipating the consultant's invasion arouses new fears. Unbounded optimism about interaction with a consultant postpones taking responsibility and making difficult decisions, but eventual recognition of real expectations raises anxieties about failure to develop real proposals.

Finally the unconscious assumptions work against a client's empowering an advisor to assist with problems. In different ways, all the assumptions portray an advisor as someone who need not act, but should only be present. None of these assumptions requires a client to acknowledge problems, share information, or develop proposals with a consultant. In the end, directly or indirectly, these assumptions lead the client to attack the consultant.

Thus both the dynamics of bureaucratic responsibility and tensions within advising relationships may unconsciously lead bureaucratic workers to want to weaken or defeat their consultants. This analysis offers psychological hypotheses for why Simon's clients might work against him, in conflict with their espoused intentions. The next section turns to anthropological studies of ritual scapegoating, to identify social characteristics of an organization which may encourage efforts to cripple advisors. The anthropological material offers evidence about unconscious rationalizations for harming advisors—they may be blamed for social problems.

SCAPEGOATING RITUAL

Anthropologists emphasize the purposes or functions of ritual activity, rather than any inherent characteristics of ritual actions. For example, the use of physical objects, sexuality, music, or sacrifice is less important than what ritual performers believe they accomplish through their actions. Geertz (1973) summarizes many anthropologists' views by describing rituals as systems of significant symbols that demonstrate the meaning of mundane events. Ritual activity is consecrated behavior, invoking and affirming the "really real" and joining it to the realities of everyday life. Ritual accomplishes this not simply by establishing logical meanings for human activities, but also by exposing participants' feelings about their activities and somehow reconciling and satisfying these feelings.

Rituals are social activity. Cultural norms prescribe when and how they should be performed, and all members of a society take part in the same or related rituals. Thus the meaning which rituals affirm is social—they demonstrate not simply that everyday events have an abiding meaning, but also that the group possesses or embodies that meaning. Freud (1913, rep. 1950) called attention to similarities between cultural rituals and individual obsessional activities, and he suggested that both respond to guilt anxiety. But individual "ritualistic" activities differ from cultural rituals in two ways. First, the problems to which the former respond are idiosyncratically perceived, whereas the latter respond to socially defined difficulties. Second, consistently, the performance of individual obsessional actions tends to isolate individuals from one another, whereas participation in common social rituals unites people (Kafka, 1983; Roheim, 1943, rep. 1971).[16]

Van Gennep (1908, rep. 1960) wrote specifically of rites of passage, but his descriptions emphasize a quality common to virtually all rituals—they enable participants to pass from one condition into another. More generally, rituals permit their celebrants to move from a state not (or no longer) appropriate into another that is. Freud (1913, rep. 1950) pointed out the individual rewards of such a transition— anxiety associated with one position is replaced by the relative security of another. Roheim (1943, rep. 1971) drew attention to the social benefits—contagious, widely felt anxiety about social survival is replaced by reassurance about the society's future. Girard (1977) suggests that an underlying purpose of many rituals is to enable members of a society to pass from a condition where reciprocal violence, unending revenge, and a "war of all against all" threaten both individual and social survival to a condition of consensus based on unanimous participation in the violent sacrifice of a surrogate victim.

Winnicott (1971) speaks of both the transitional quality of cultural activity and its mediation of individual relations with society. He argues that cultural activities, including rituals, grow out of childhood play and are similarly concerned with mediating anxiety-ridden relations between individuals and the social and physical environments. Cultural rituals differ from child's play in being prescribed by an inherited tradition. Within a tradition, an individual may innovate, play, and create new forms, although rituals remain relatively fixed. Thus rituals may be regarded as socially sanctioned play. They help individuals feel more secure about their separations and differences from others. At the same time, they help secure the continuance of a society in which individuals are separate and different.

Rituals have four basic characteristics. First, they express normally hidden feelings about the meanings of everyday activities. Accordingly, second, the "logic" of rituals is not conventional reasoning, but, at least largely, a logic of the unconscious mind. Third, the problems to which rituals respond, as well as the ritual procedures themselves, are socially defined. This means that the problems are social rather than individual, and that the remedies reaffirm the society, even at individual cost. Fourth, ritual, like child's play, provides participants with security from anxiety.

THE SCAPEGOAT

One common ritual, which bears illuminating resemblances to Simon's consultation, is the creation and sacrifice of a scapegoat. Societies create scapegoats as vehicles for the expulsion of sins. Members of a society symbolically transfer their sins to a scapegoat, whom they then sacrifice as a means of absolving themselves of guilt. In selecting a victim, members of a community choose someone who is simultaneously similar to and different from themselves. The similarity makes the transfer of sins credible, and the difference makes the sacrifice acceptable. A successful scapegoat ritual serves two purposes. Morally, it purifies the community of sins. At the same time, in removing from the community any actions that might offend members, it eliminates causes for revenge and sources of strife, thus bolstering social order and harmony.[17]

Frazer's (1940) studies of magic and religion throw light on typical characteristics of scapegoating. Two examples illustrate its basic dimensions. The first comes from an African tribe:

> At Onitsha, on the Niger, two human beings used to be annually sacrificed to take away the sins of the land. The victims were purchased by public subscription. All persons who, during the past year, had fallen into gross sins, such as incendiarism, theft, adultery, witchcraft, and so forth, were expected to contribute 28 ngugas, or a little over £2. The money thus collected was taken into the interior of the country and expended in the purchase of two sickly persons "to be offered as a sacrifice for all these abominable crimes—one for the land and one for the river." A man from a neighboring town was

hired to put them to death. On the twenty-seventh of February 1858 the Rev. J. C. Taylor witnessed the sacrifice of one of these victims. The sufferer was a woman, about nineteen or twenty years of age. They dragged her alive along the ground face downwards, from the king's house to the river, a distance of two miles, the crowds who accompanied her crying, "Wickedness! Wickedness!" The intention was "to take away the iniquities of the land. The body was dragged along in a merciless manner, as if the weight of all their wickedness was thus carried away." (Frazer 1940, pp. 569–570)

A second example comes from India and Albania:

In November the Gonds of India worship Ghansyam Deo, the protector of the crops, and at the festival the god himself is said to descend on the head of one of the worshippers, who is suddenly seized with a kind of fit and, after staggering about, rushes off into the jungle, where it is believed that, if left to himself, he would die mad. However, they bring him back, but he does not recover his senses for one or two days. The people think that one man is thus singled out as a scapegoat for the sins of the rest of the village. In the temple of the Moon the Albanians of the Eastern Caucasus kept a number of sacred slaves, of whom many were inspired and prophesied. When one of these men exhibited more than usual symptoms of inspiration or insanity, and wandered solitary up and down the woods, like the Gond in the jungle, the high priest had him bound with a sacred chain and maintained him in luxury for a year. At the end of the year he was anointed with unguents and led forth to be sacrificed. A man whose business it was to slay these human victims and to whom practice had given dexterity, advanced from the crowd and thrust a sacred spear into the victim's side, piercing his heart. From the manner in which the slain man fell, omens were drawn as to the welfare of the commonwealth. Then the body was carried to a certain spot where all the people stood upon it as a purificatory ceremony. (Frazer 1940, pp. 571–572)

Several themes are common in these examples. The first is the explicit connection between the sins or offenses of the mass of people and the punishment of the scapegoats as symbolic repositories of those sins. At the same time, however, responsibility for the murder of the victims is ambiguous. For example, although at Onitsha the future victims are purchased by public subscription, the identity of the individuals acquired is determined by their sickliness, which already marks them. In other cultures, the victims are ugly or deformed persons, often kept in ready captivity at public expense. Sickliness, ugliness, and deformity are characteristics visited on people by a divine power. Those who select the victims are only observing the divine designations. Those who subsequently kill the victim only take someone indicated by the gods for sacrifice. Thus it is possible to kill a scapegoat as a way of expiating one's own sins without being directly responsible for the murder.

The role of divine power in identifying the victim also beclouds the relationship between the victim and the mass of people. In these examples the victim is human. This essential identity between the victim and the people makes transference of sins believeable—and thus effective. And yet each victim is also different from other people. At Onitsha the victim is inferior in health, although this condition is peculiarly compensated by the implication that it is divinely conferred. This same ambiguity is found in India and Albania. The victim among the Gonds is struck by divine madness, rendered both less and more than human. In Albania the victims are either insane or inspired and, corresponding to this duality, are kept in the ambiguous position of sacred slaves. The difference from other humans suggests that the scapegoat is already separate from human society and permits corporeal sacrifice.

The description of the Albanians is particularly explicit about the divine characteristics of the victim. This is someone who is inspired by extraordinary knowledge and can even prophesy the future. In all cases, by the time the victim is ready to be killed, he or she has come to "know" the sins of the people, an extraordinary feat of comprehension. Sometimes the victim's acquisition of this "knowledge" is formally enacted; in some cultures the scapegoat is required to engage in sexual and other forms of license for a fixed period of time just prior to the sacrifice. Finally, the divine power of the scapegoat rests centrally in the victim's ability, in fact, to absorb the sins of others and to absolve them by dying. That is no ordinary human accomplishment.

Still, there is something puzzling about the divinity of the victim, particularly when it is so explicit as among the Gonds and Albanians. How is it possible to kill a god, or a god's representative? And how could it be permissible? Two different hypotheses both emphasize the benefits of sacrifice to the victim as well as to the community. Each regards the scapegoat as a dying god but emphasize a different element in this dual identity. Frazer emphasizes the ongoing divinity of the victim. He suggests that the scapegoat may be perceived as a dying god whose sacrifice may be considered not simply aggression, but also succor:

> If we ask why a dying god should be chosen to take upon himself and carry away the sins and sorrows of the people, it may be suggested that in the practice of using the divinity as a scapegoat we have a combination of two customs which were at one time distinct and independent. On the one hand we have seen that it has been customary to kill the human or animal god in order to save his divine life from being weakened by the inroads of age. On the other we have seen that it has been customary to have a general expulsion of evils and sins once a year. Now, if it occurred to people to combine these two customs, the result would be the employment of the dying god as a scapegoat. He was killed, not originally to take away sin, but to save the divine life from the degeneracy of old age; but, since he had to be killed at any rate, people may have thought that they might as well seize the opportunity to lay upon him the burden of their sufferings and sins, in order that he might bear it away with him to the unknown world beyond the grave. (Frazer 1940, pp. 576–577)

In this view, killing the god figure is not murder, since the god is dying anyway. Indeed, without the sacrifice, the god would die without dignity.

Girard (1977) suggests, alternatively, that incidentally selected victims, who are certain to die, come to be seen as gods or godlike only in relation to their anticipated demise. In other words, people's belief that the sacrifice of their victim absolves them of sins and violent impulses toward one another leads them to regard such an apparently powerful person as divine. Thus the victim becomes divine by dying in a successful sacrifice, and this attribution of divinity is a gift to an otherwise inconsequential person.

While these two views regard the origins of the scapegoat's divinity differently, they converge in one paradox. The victim is an extraordinary being, but the height of his power is revealed only in his sacrifice. Moreover, the community is commanded to sacrifice the victim, who benefits from being killed.

Still, even though these ideas give some intellectual logic to the sacrifice of divine figures, it would be surprising if the accompanying emotional state were not highly ambivalent. The sacrifice of a god is presumptuous, and apprehensions of revenge must persist. Indeed, rituals accompanying the killing of victims express both aggression and propitiation. At Onitsha the selection of the future victim begins with the payment of a fee, an indication of good faith toward the victim. India and Albania are typical in maintaining the future victim in luxury, feeding the victim amply, anointing the victim with oils, or dressing the victim in priestly raiments. These are not simply signs of respect; one must speculate that they are anticipatory recompense. Indeed, among many peoples who slayed animals and feared revenge from the victims' spirits or survivors, it was common to mollify the future victim with excuses, flattery, food, and gifts.

With human scapegoats the killing might be surrounded by acts of torture or dismemberment. The ancient Greeks beat their victims upon the genitals with squills and branches of certain trees. The Yoruba beheaded their victims. Others removed such organs as the liver, ears, skin of the forehead, or testicles and incorporated them into a ritual meal. Certainly these physical acts, committed on human victims, are aggressive. Yet, Frazer argues, they were also seen as acts of respect, recognizing the supernatural powers of the victims and seeking to incorporate them into the living. Thus the organs may be regarded as the seats of divine understanding, courage, perseverance, strength, or fertility. In this way the dying god is preserved from an ignominious death and is kept alive symbolically by transfer of the divine attributes to the masses. Not only do the members of society expiate their sins by sacrificing the scapegoat, but the people gain new virtues as well.

SCAPEGOATING THE ADVISOR

Frazer's accounts show that scapegoating is widespread. Moreover, his records of practices in traditional societies are instructive because of the seemingly direct relationships between unconscious concerns and overt actions. These societies acknowledge a belief in magic and embrace rituals that express it, whereas contemporary Western society is likely to conceal magical beliefs and their unconscious connections beneath scientific rationalizations. Although there are dangers in drawing inferences from practices in one culture to another, Frazer's examples suggest analogues to traditional scapegoating rituals in contemporary consultation. While the physical treatment of the scapegoats whom Frazer describes never appears in relations between bureaucratic workers and their advisors, underlying psychological dynamics have apparent counterparts.[18] How accidental is it that clients often tell consultants that "we want to pick your brain"?[19]

The success of much traditional scapegoating depends on the perception of the scapegoat as both similar to and different from the people whose sins are to be expiated. In bureaucratic organizations, an advisor, ambiguously both inside and outside the organization, satisfies these conditions when organizational wrongdoing must be remedied. Specific steps in the preparation of an advisor for the role of scapegoat turn on alternately emphasizing the advisor's identity with the client and the advisor's deviance. In this process, as the next chapter reveals, an advisor may be a more or less willing participant.

The following discussion describes how bureaucrats may create potential scapegoats by shifting blame for problems onto advisors, and then, how workers may symbolically sacrifice their scapegoats by harming the advisors.

Shifting Blame for Problems

The advisor's ambiguous position as invited intruder presents a problem: here is someone who cannot be accounted for according to the formal structure of the organization. One response may be to reduce the intrusion by strengthening the invitation to be a member of the organization. Thus the client may attempt to lure the advisor into surrendering distance on the organization and its problem and joining with the client. Seductive efforts include praise of the advisor's talents along with hints that the advisor has worked sufficiently long with the client to have gained a special understanding of and sympathy for the client. The advisor should recognize that the client's point of view is only reasonable, given the complexities of the bureaucracy, its occasional craziness, the difficulties of doing competent work, and, surely, the client's good intentions. Such a seductive effort may be intended to pacify suspected hostility from the advisor in order to get a supportive consultation. It also attempts to secure a tacit pledge of brotherhood (or sisterhood), a strong tie of loyalty which would rule out "telling tales outside the family."[20]

And yet the distance of the advisor is irreducible—the advisor really is an outsider and can be effective only so long as he or she remains so. Hence a client may attempt to reduce the ambiguity of the advisor's status alternatively—perhaps simultaneously—by emphasizing the intrusiveness and exaggerating the distance. A client may take steps—for example, withholding information—in order to keep the advisor outside the organization. More intricately, the client may deliberately encourage the consultant to act in ways that appear to be voluntary deviance.

For example, once the advisor knows the problems of a worker who feels anxious before hierarchical superiors, the worker may attempt to give the advisor knowledge of the problems or offenses of those superiors as well. Flattering the advisor's special understanding may be the seductive prelude in this peculiar courtship. Here the advisor is tacitly encouraged to discover and utter ineffable organizational secrets. Even if the client withholds some information from the advisor, he or she may also give the consultant information about these secrets, which may be passed on in a way not attributable to the client. The client may threaten to deny any knowledge if it is ever publicly discussed, thus increasing the advisor's estimation of the information and the likelihood of revealing it. Less provocatively, the client may simply arrange for the advisor to "discover" the same information.

An advisor may be particularly receptive to hints of organizational secrets because of experience that such secrets do, in fact, impede problem solving. Nevertheless, the divulgence of these secrets is an act of hostility. While the client's seduction draws the advisor in, encouragement to utter undiscussable secrets and to make statements about the actions of superiors puts the advisor beyond the organizational pale. No one who utters organizational secrets may be associated with the organization.

Crucially, members of the organization experience the advisor's naming of secret offenses as if the advisor has taken on those very offenses. Even if, at least unconsciously, people recognize the truth of what the advisor says, overtly they may deny that such things as selfishness, ruthlessness, duplicity, or deceit exist in the organization. Because the advisor is the only one who identifies such offenses, surely the consultant "creates" them by mentioning them, and they might be eliminated by expelling the advisor.[21] By naming these offenses, the advisor takes an important step toward becoming a scapegoat. The expulsion of the advisor will defend the members of the organization from these offenses.[22]

The advisor's apparently voluntary deviance from organizational norms resembles the ugliness, sickliness, and deformity of the scapegoats in Frazer's accounts. These are all stigmata settled on future victims by someone other than the executioner. They make it possible to harm the scapegoat without responsibility for the injury. Perceptions of advisors especially resemble the Albanians' perceptions of their sacred slaves. They know and say things which ordinary humans do not recognize. Yet they are regarded as ambiguously inspired or insane. When the advisor utters an undiscussable organizational secret, it can be recognized as both an extraordinary insight and the blathering of an ignoramus. Although the apparent "insanity" of the advisor mitigates any guilt from doing harm to the advisor, the appreciation of the advisor's distinctive intuition and knowledge justifies a magical belief that the client and others will absorb the advisor's knowledge by doing away with the advisor.

Harming the Advisor

The advisor's punishment for organizational offenses is to be rendered ineffectual. Clients cripple their consultants in ways that seem to eliminate the anxiety of advice seeking, exculpate the clients from responsibility for problems, and restore at least a modicum of organizational tranquillity.

Withholding Information

The first step in disabling an advisor is to withhold information about the problem. A worker may do this in order to avoid shame about having a problem or punishment for poor judgment. The client may neglect to give the advisor certain information, not tell the advisor about particular sources of information, or directly deny the advisor access to specific sources. The worker may feel there are limits to the degree a consultant may delve into the life of an organization and may justify denying access on the grounds that the information is not pertinent to the advisor's work and that disclosure would violate the ethics of confidentiality.

The advisor's position as invited intruder provides justification for withholding information. Even though some argue that boundary spanning is important for solving problems (see Bennis & Slater, 1968; Biller, 1973; Michael, 1973), the advisor's position threatens the security of organizational boundaries. Any action the consultant would take or recommend overtly endangers existing limits on individual or organizational responsibility. Workers may respond by emphasizing traditional boundaries and shoring up defenses against the intruder. Regardless of misgivings about organizational practices or animosity toward organizational members, a bureaucratic worker is likely to find loyalty to the organization more reassuring then following an outsider. A worker who unconsciously associates the consultant's calculations with an evil father's intrusions into a nurturant maternal realm will become still more anxious about organizational boundaries and more determined to withhold potentially treasonous information.[23]

Resentment about dependency on an advisor may reinforce these impulses. Insofar as transference of childhood feelings onto the advisor stimulates feelings of incompetence, an adult worker may withhold information as part of an unconscious strategy to punish the advisor and make him or her as impotent as the worker feels. The client may rationalize this action with the thought that, if the advisor were sufficiently competent to justify this dependency, then the advisor could get all the necessary information alone.

Discrediting the Advisor's Competence

Similar motivations may lead to a second step in disabling the advisor, discrediting the advisor's competence. If the advisor can be found to be incompetent, then it is neither necessary nor appropriate to share information with him or her, and it would be foolish to accept his or her recommendations. Again the ambiguity of the consultant's role provides a framework and justification. A client may attempt to reduce this ambiguity by exaggerating the advisor's deviance. The client may identify aspects of the problem—or, more generally, the organization—that the advisor cannot understand and about which is unable to make any judgment. For example, it can be said that the advisor does not know the city (or the neighborhood, the county, or the region) very well, that the problem is affected by complex laws and regulations about which the advisor knows little, or that the problem belongs to a field of expertise in which the advisor has little training. These deficiencies may be weighed more heavily than all the advisor's training in judging him or her incompetent to handle the problem at issue.

Discrediting a consultant satisfies a number of needs. If the advisor is not a competent observer, then any errors the advisor may discover and attribute to a bureaucrat may be ignored as misunderstanding. In addition, shame about having a problem is removed, although paradoxically. Initially an advisor may be employed because he or she is considered to be the most competent assistant available. However, if the advisor is subsequently discredited, then this incompetent person's advice is worthless and need not be followed. Hence if even the (initially) most competent advisor cannot offer any useful advice in solving the bureaucrat's problem, then the worker need feel no shame about having the problem, because it is insoluble and, by implication, excusable. The discrediting of the advisor vindicates the bureaucrat.

Further, the devaluation of the advisor's expertise releases the worker from taking any action on the basis of the advisor's word. If the worker perceives him- or herself in a situation where responsibility without authority poses the risk of being scapegoated for failures, the discrediting of the advisor provides a temporary reprieve from taking action for which the worker might be held culpable. Finally, discrediting the advisor's competence rescues the worker from the humiliation of a dependency relationship. If the advisor, after all, is not highly competent, then the worker has no reason to be dependent on the advisor and can treat the advisor in any manner whatsoever.

Sabotaging Recommended Solutions

The final step in disabling an advisor is to sabotage any recommendations for solving problems. A failure to solve the bureaucrat's problem is a demonstration of the advisor's incompetence. Motivations for sabotage and its rewards are the same as those encouraging other attacks on the advisor.

Again, the advisor's status as intrusive outsider provides a framework and justification. Not only is the advisor a potential disruptor of organizational relations, but as an outsider, he or she has no direct stakes in change and does not share any of the risks of action. Sabotaging recommendations may both protect the bureaucrat and punish the advisor. Thus a bureaucrat, citing an intimate understanding of the history and practices of the organization, may find many reasons why any recommendation either will not succeed in solving the problem or would be dangerously disruptive to normal organizational procedures. If the client does act on the recommendations, he or she may implement them poorly, perhaps rigidly following their letter while overlooking their spirit. Or the worker may give lip service to the recommendations while continuing to follow traditional, even if problematic, modes of action. Or the client may simply ignore the recommendations.

The result of any of these actions is that there is no opportunity to put the advisor's recommendations into practice and discover that they may solve a problem which the client could not. In this way the organization is exculpated, and past practices are vindicated. The individual bureaucrat is saved from shame at having a problem, as well as culpability for any perceived errors. Finally, insofar as the advisor fails at solving the worker's problem, then the client has no further reason to feel dependent on the advisor.[24]

The Advisor Both Blamed and Harmed

The processes of blaming and harming the advisor are interlaced. For example, a client may at one moment discredit an advisor's competence to agency colleagues, shortly thereafter confide organizational secrets to the advisor, at the same time withhold historical organizational data from the advisor, later publicly laud the advisor's recommendations, and finally, continue with traditional programs. The combination of these actions has the cumulative effect of shifting blame for organizational problems to the advisor and punishing him or her for those problems.

Often the advisor's departure is regarded with relief. Ostensibly the relief reflects the advisor's success in solving a problem. Admittedly, the relief often follows stressful efforts to identify, discuss, and resolve complex issues. Unconsciously, however, organizational members feel relieved because they believe that their treatment of the advisor has absolved them of shame or guilt for their problems. Further, the workers feel strengthened by what they have taken from the advisor.

The Consultation of the Three Incidents

These analogies to traditional scapegoating in contemporary advising offer clues to the puzzles posed by the three incidents in Simon's case, even though more information about the consultation would be necessary for testing these interpretations.[25] First, why would a client go through the effort of seducing an advisor into an alliance and then deliver the advisor up to punishment? At least unconsciously, the assistant director (who, it must be recalled, was deputy to the formal client) was attempting to solve organizational problems by scapegoating the advisor.

The first incident, in which the consultant was likened to Jose Ferrer as Cyrano de Bergerac, served two purposes. Overtly, the assistant director praised the advisor's appearance and insight, encouraging him to think of himself as her ally. At the same time, the reference to Cyrano at least unconsciously marked the advisor as ugly and separate. This differentiation paralleled the reality that the members of the client group were black, while the advisor was white. Thus the advisor, like a typical scapegoat, was considered simultaneously an exceptional insider and an ugly outsider.

The second incident, in which the advisor presented criticisms of the director at a meeting which the director attended, identified the advisor with the scapegoat role. The assistant director's continuing flattery of the advisor offered a gift to the future victim. Her and the coordinator's encouragement to report criticisms of the director amounted to a symbolic imparting of organizational "secrets," or sins. The consultant knew that the director intimidated and humiliated subordinates, who then became angry but felt unable to protest for fear of retribution, and he knew that this shame and anger contaminated most organizational relationships. His recitation of these criticisms at the meeting completed the transmittal of the organization's sins to him by publicly identifying the consultant with these sins. The appearance of the director foreshadowed the ultimate sacrifice of the advisor.

The third incident, in which the assistant director revealed an earlier consultant's work, threatened the advisor with death. Her actions told the advisor that the client had the power to withhold essential information, had already done so, and might do so again. In case the practical consequences of these possibilities failed to impress the advisor, the specific information revealed—the firing of a similar previous consultant—foretold the inevitable death of this consultant.

The conclusion of the consultation marked the advisor's death. The celebratory reception of the report expressed both gratitude for receipt of his strength and relief at his anticipated departure. Once he was gone, there would be no risk that anyone would mention organizational strife, and the report gave new force to a presumably more unified effort to move ahead. Tacitly, members of the director's staff and council members could unite by redirecting any internal hostility into a unanimous, albeit appreciative, dismissal of the consultant.[26] No one seriously attempted to keep him alive by trying to implement his recommendations. When the coordinator did initiate experimental program modifications, he delivered the finishing blow—citing the advisor's name and report but ignoring the substance of his advice. The advisor was dead; long might his clients live.

WHY PUNISH THE ADVISOR?

Still there is one other, perhaps obvious, puzzle remaining from Simon's consultation. Even though the scapegoat drama has a compelling logic, its "solution" of organizational problems is only magical. The original problems remain. In the case of the council program, unequal distributions of responsibility and authority and (now under a new director) an aimless but potentially punitive management style continued to paralyze hundreds of county workers in biweekly meetings. The answer lies in the fact that the bureaucratic psychological structure may make the problems which scapegoating an advisor can solve seem more compelling than the organizational problems for which consultation is requested.

When bureaucratic predicaments make workers anxious before superiors, subordinates may be expected to enlist others, such as consultants, in their cause. Clients may assign advisors the part of a double-agent in an unconscious three-act drama. This play turns on the clients' projecting their anger onto the advisors, that is, unconsciously attributing their own anger to the advisors and encouraging them to express it.

In the first act the client recruits the advisor against the client's superior, to whom the client has previously attributed his or her own punitive impulses, as described in the second chapter. This alliance is evident in the first incident of Simon's consultation. The association protects the client because, insofar as the advisor expresses the client's hostility, the client may deny any harmful wishes. Further, in any conflict, the advisor is the one who will be wounded.

Simon's second incident reveals the advisor acting on the client's anger, but also hints at the client's efforts to harm the advisor, revealed more clearly in the third incident. A hint at the reason for this puzzling turnabout is found in the client's early profuse flattery of the advisor. Although this flattery cements the alliance, it may also serve to propitiate the advisor for any anger against the client. But why should an advisor be hostile toward the client? In one way, this fear may seem reasonable, in that a consultant, aroused to anger by a client against a superior, might turn against others as well, including the client. However, the client's anticipation of the consultant's hostility is more specific.

The advisor's aggression is the theme of a second act, in which the client unconsciously aligns the consultant with bureaucratic superordinates. This alliance is based on transference similar to that which identified the superior with a punitive parent earlier. As noted, dependency on an advisor may stimulate resentment. In turn, a client may unconsciously deny any negative feelings and imagine, instead, that any anger in the consulting relationship belongs to the advisor. At the same time, in searching for models from whom to draw inferences about the advisor's feelings and motives, a bureaucrat may transfer assumptions from the organization's nearby top administrators. Thus a client may unconsciously come to see the consultant as similar to a bureaucratic superior and allied with that person. Insofar as the worker imagines the administrator to be an instrument of conscience, the worker may assume likewise of the advisor, expecting him or her to punish the client for hostility toward the boss. Flattery may be an effort to buy off conscience.

The client attempts to avoid this punishment in a third act by forming a new unconscious alliance, with the superior, against the advisor. Recognizing ongoing dependence on organizational superordinates, the client tacitly reassigns hostility and virtue in a way that makes this reality bearable. Thus the client chooses to regard the superior as reasonable and good and recalls that the advisor all along has been the agent of the subordinate's destructive aims. In this context, ruin of the advisor may now be regarded as prudent. The client must preemptively strike the consultant or be treated far worse—and legitimately—by wounded bureaucratic superiors. Most generally—and this is the abiding theme of the scapegoat drama—the client may align him- or herself with others in the organization in attributing all antisocial impulses to the consultant, who then must certainly be punished.[27]

This psychic drama may help a bureaucrat avoid shame and guilt anxiety aroused by organizational dependency. It aids subordinates contemplating aggression against hierarchical superiors whose actions or positions shame them. The first act permits at least limited, vicarious expression of hostility toward organizational superiors; complete success is inevitably frustrated by fears of being fired. However, the third act, in which subordinates succeed in unconsciously redirecting hostility from bosses toward a consultant offers more satisfying opportunities for aggression. Workers can repeatedly strike someone who cannot punish them, while denying any hostility toward the bosses, thus at least temporarily freeing themselves from anxiety.[28]

Unconsciously, a bureaucrat may believe that the process of blaming, harming, and banishing an advisor may magically solve organizational problems. Here is an answer to the second puzzle posed by Simon's consultation. As shame and guilt anxiety increase, a worker may gradually feel that they are more important, because more painful, than formal organizational problems and give priority to remedies for anxiety. Thus magical solutions seem efficacious in solving overt problems when they relieve the worker of anxiety associated with them. Hence a worker may prefer the magical process of scapegoating to instrumental attacks on problems. In this way, a subordinate may continue to work in an organization that shames without suffering adverse consequences for reacting to shame anxiety.

Accounts of traditional scapegoating emphasize the collective benefits of the ritual. Frazer, for example, observes that the scapegoat is often punished for others' sins only after being required to act them out. The Roman Saturnalia is the most vivid example of mass license, followed by the sacrifice of a victim identified with Saturn. Such rituals express resentments against the burdens of civilization and satisfy wishes to violate social norms. Crucially, the restoration of order exacts just punishment for moral violations and reasserts secure social relations. Thus society is reaffirmed, or recreated. Girard emphasizes that the restoration of order rests essentially on a collective redirection of internal violence outward toward a unanimously designated substitute victim. Consequently, harmonious relations and cooperative endeavors become possible.

Bureaucratic subordinates' aggressive wishes toward their superiors present analogous threats to the stability of the organization. Scapegoating an advisor provides two outlets for rebellion. By leading a consultant to act out hostility toward administrators, subordinates may vicariously rebel against constraints. Then, by punishing the consultant, organization members may simultaneously freely express their aggressive wishes toward a surrogate victim and pull together in apparently renewed unison. The advisor's removal provides both required atonement for insubordination and a reassertion of the organization.[29]

There are risks in drawing straightforward analogies from rituals in traditional societies to rituals in formal organizations. Various writers (for example, Toennies, 1887, rep. 1957, Weber, 1921; rep. 1964; Weick, 1979) have emphasized that bureaucracy differs from traditional societies in that its membership is partial, temporary, and more specialized. Members can leave the organization and even nurture fantasies of defeating it, as Chapter 5 suggested. Hence these observers conclude that members' emotional loyalty to contemporary organizations is less total and intense than that of members of traditional societies. Contemporary institutional specialization certainly does divide emotional investments, but whether investment in any one institution is less intense than in traditional societies is hard to determine. At the least, evidence in this book demonstrates that members feel strongly about the operations and rewards of their work organizations.

Even so, many observers continue, contemporary members of formal organizations recognize the differences between magical action and instrumental, scientifically based action. At best, the former may have something to do with "symbolic" issues, morale, or workers' feelings, but modern managers and workers understand that they have come together to deal with instrumental tasks. Yet the body of organization theory may be read as a debate over how to organize workers so as to balance instrumental goals (maximizing productive efficiency) and "symbolic" goals (maintaining worker loyalty) (see Denhardt, 1984; Hummel, 1982). Disagreements concern how much workers' needs conflict with efficient production and how much efficiency must be sacrificed in order to satisfy how much of workers' needs in order to maintain a work organization.

Managers' stakes in preserving their organizations while seeking efficiency are obvious. Crucially, examples in this book demonstrate that subordinates also intimately understand connections between organizational harmony and competent work. While efforts to establish satisfying psychological contracts lead to assertiveness and rebellion against bureaucratic constraints, these efforts are above all a search for supportive organizational work conditions. Even when workers attempt to escape their organization, they reject it in favor of a more fulfilling organization. At least unconsciously, workers recognize that they can solve problems effectively only when they can collaborate with others in powerful ways. Many unconsciously motivated actions, such as scapegoating, may be interpreted as magical efforts to maintain—or more likely create—organizations that promote competent work.

Nevertheless, scapegoating fails to solve the instrumental problems with which bureaucracy is formally charged. Perhaps workers' invocation of a magical realm in which scapegoating "works" evokes fantasies of a primeval organization which satisfies their interests in competence and recognition. However, emphasizing magical solutions deprives them of opportunities to test their real ability and fashion solutions to real problems. The organization may survive, but it is not competent. Hence one may speculate that a subordinate's preference for magic over instrumentality is also an unconscious effort to undo the organization, because avoiding instrumental action leaves its formal problems unsolved. Thus a subordinate may scapegoat an advisor not simply to relieve anxiety, but also to undo, or cripple, the organization that arouses shame to begin with. This revenge may be the ritual's strongest attraction of all.

CONCLUSION

Despite all the talk of hostility, aggression, and violence in relation to advisors, the fact remains that Simon's consultation proceeded more or less conventionally. He might have been more successful if his client, the director, had not switched jobs. No crisis ever occurred, although scapegoating may become explosive. In this case, the advisor's "expulsion," or "death," took the banal form of thanking but ignoring him. Still, bureaucratic workers may harm themselves by resisting advice, and it is important to understand when they may unconsciously choose to do so.

Many bureaucratic workers effectively use formal and informal consultants to solve organizational problems. Although common dynamics of organizational responsibility and tensions in consulting relationships may contribute to resistance, it is not inevitable. The dynamics of organizational responsibility are most likely to lead to scapegoating an advisor when (1) organizational tasks or problems are ambiguous; (2) organizational members are locked in conflict with one another, whether over personality, procedural, or substantive issues; or (3) the client feels especially ambivalent about asserting him- or herself, either because of general organizational dynamics or because of personality idiosyncrasies. Tensions in consultation are most likely to lead to scapegoating an advisor when (1) the advisor is not familiar to the client (although familiarity does not guarantee undistorted conscious communication); (2) the individual client or the organization lacks well-established advice-seeking procedures (although established procedures are not necessarily conducive to undistorted communication); or (3) the advisor acts in ways that elicit mistrust or unconscious transferences—for example, by acting parentally or condescendingly or by expressing ambivalence about where he or she stands in relation to the client and the organization.

These conditions are not uncommon, and Simon's consultation, by demonstrating how easy it is for bureaucrats to resist advice, shows the difficulty and complexity of many advising relationships.

NOTES

1. For a discussion of the varieties of consultation requests and responses, see Lippitt and Lippitt (1978).

2. For an analysis of variations in clients' requests and styles of working with consultants, see Meltsner (1976).

3. See Baum (1983a) for a model of multidimensional leadership appropriate to collaboration between professional planners and lay citizens.

4. See Benveniste (1977) for numerous examples.

5. Lasswell's (1960, 1963) seminal books are exceptional among political scientists in considering the combination of conscious and unconscious interests motivating administrators and other political actors.

6. Forester (1982, 1984) offers a framework for examining sources of distortion in communication among planners, administrators, politicians, and others who participate in social problem solving. Although he recognizes unconscious actions, the framework emphasizes politically motivated distortions for two reasons. First, they may be visible in both their calculations and their consequences. Second, once identified, the sources of many of these distortions may be removed, and these changes would improve communication significantly. This book holds that, in addition, some important sources of distortion stem from unconscious interests and that these interests may also reinforce politically motivated distortions.

7. Argyris and Schön (1974, 1978) argue that defensiveness in advising relationships is universal, or at least endemic to industrial societies (Argyris, 1982). They argue that bureaucracy only intensifies general defensive tendencies.

8. Bonazzi (1983) distinguishes "expressive" and "instrumental" scapegoating, corresponding to the distinction between unconscious and "political" motivations to scapegoat. However, he concedes that the boundary between the two categories is fuzzy.

9. For case examples of ways in which organizational members adopt and adhere to ssumptions that conflict with their expressed purposes, see Argyris and Schön (1978), Crozier (1964), Gouldner (1965), Menzies (1975), and Rice (1963). Argyris and Schön provide numerous examples of workers' insistence on keeping such assumptions undiscussable "secrets."

10. I am grateful to Natalie Cantor for bringing this example to my attention.

11. A male advisor may also be perceived as a fraternal or avuncular figure. A female advisor is likely to be perceived as a maternal figure. The sex of the advisor does not fully determine the content of the unconscious transference. Other influences are the relative ages

of the advisor and client, as well as their personalities. This chapter emphasizes the dynamics of a paternal transference, although other perceptions are possible. For a discussion of the varieties of transference in organizations, see Hodgson, Levinson, and Zaleznik (1965); and Kets de Vries and Miller (1984).

12. For a discussion of resistance, see Brenner (1976), Fenichel (1945), and Freud (1946).

13. Michael (1973) discusses the ways in which organizational and professional norms discourage an acknowledgment of the need for an advisor.

14. Lippitt and Lippitt (1978) identify all these resistances to consultation from their own practice:

> [A] serious and pervasive block [to requesting advice] is the evaluation that to "do it yourself" is the greatest sign of competence and asking for help is a symptom of weakness. Because the highest value is placed on the line people who are the producers, there is a reluctance to invest in staff people, such as internal consultants. If help is needed, there is a tendency to depend on inside resources in order to "keep it in the family." Often, the specific type of help needed to work on particular problems is not available within the organization, but people are reluctant or lack sophistication on how to seek the appropriate outside help. (p. 4)

15. For a discussion of how these unconscious assumptions arise in groups, see Bion (1961) and Slater (1966). For an examination of how these assumptions may dominate organizational life, see Kets de Vries and Miller (1984). For analysis of how these assumptions may enter into and affect relations between planners or architects and their clients, see Bexton (1975).

16. There is a well-known anthropological dispute over whether participants regard ritual activity as necessary to prevent certain disasters, or whether participants regard it as a remedy for the occurrence of disasters. Radcliffe-Brown (1939; rep. 1965) took the former view, while Malinowski (1931; rep. 1965) held the latter interpretation. Geertz (1973) suggests that their disagreement really stems from different perspectives on ritual performance. Radcliffe-Brown, thinking sociologically, emphasized the social meaning of ritual. He observed that members of societies performed rituals in order to head off certain impending troubles and to reaffirm the societies' meaning. Malinowski, thinking social psychologically, emphasized the individual's interpretation of his or her participation in a socially prescribed ritual. Malinowski observed that individuals performed rituals when taboos or other norms had been violated, in order to repair any damage, and again, to reaffirm the society that held the norms sacred.

17. Girard (1977) argues that the primary purpose of sacrifice is to re-establish social order. Whenever members of a society are tempted to act violently against one another, they risk setting into motion a chain of revenge that could destroy the entire community. A sacrifice provides the opportunity for all to act violently against a single agreed upon victim, who symbolically stands for all the intended victims within the community and yet stands outside the community. The importance of this difference, this symbolic position outside the community, is to identify the victim as someone on whose behalf no one would take revenge.

18. Mitroff (1983) has drawn on a Jungian framework to explore ways in which archetypes may influence organizational activities. This view would suggest that the archetype of the scapegoat has operated in both the scapegoat dramas recorded by Frazer and Girard and the advisor-client relations described in this chapter. Mitroff's discussion of the recurrence of specific archetypes in organizational life provides compelling evidence that, whatever the specific psychological dynamics involved, archetypal dramas represent basic themes in human existence. See Bettelheim (1977) for a psychoanalytic interpretation of common themes in fairy tales. See Martin (1982) and Martin et al. (1983) for an examination of typical organizational dramas.

19. Consider Frazer's (1940) report that "the Kal of German New Guinea eat the brains of the enemies they kill in order to acquire their strength" (p. 492).

20. Such a seductive maneuver may also be consciously enacted for political reasons. Reducing the advisor's critical capacity blunts any analysis.

21. This discussion points to another reason for hostility toward planners and other professional problem solvers. Because these practitioners may be the first to articulate certain problems, others unconsciously regard them as the "creators" not simply of the problem statements, but of the problems themselves, and may react with appropriate anger.

22. For an analysis of clients' simultaneous actions to seduce and create deviance in others, see Eagle and Newton (1981). See Paul and Bloom (1970) for an examination of ways in which the undiscussability of a problem gives rise to the need for a scapegoat.

23. This unconscious process is discussed by Bexton (1975). A worker may consciously conceal information from an advisor for political reasons as well. For example, the worker may seek to use an advisor's work against an adversary, concealing this tactic from the consultant. See Benveniste (1977).

24. Nevertheless, after all this has been said, it should be clear that resistance to change is not peculiar to bureaucracy, nor is all bureaucratic resistance simply psychological. Schafer (1976) reviews psychoanalytic understandings of individual resistance to change. For combined psychological and sociological analyses of organizational resistance, see Kaufman (1971), Michael (1973), Schön (1971), and Zaltman, Duncan, and Holbek (1973). Interestingly, these four books were written in the same general period, probably in response to optimism in the 1960s regarding potentials for far-reaching societal change.

25. In order to test the following interpretations, it would be necessary to discuss them with participants in Simon's consultation to see whether either their overt responses or other reactions were consistent with the interpretations.

26. The enthusiastic reception of Simon's report, including quasibiblical references to it, may have expressed the utopian beliefs of the "pairing assumption": the client may have regarded the report as the magical offspring of a relationship with the advisor. Frazer's accounts would suggest that the pleasure at the report's appearance additionally expressed pleasure at the anticipated magical devouring and incorporation of this piece of Simon into the client's organization.

27. This drama may be described in other terms. When the client assigns the advisor to attack bureaucratic authority, the advisor may be interpreted as an expression of the client's id against the client's superego. When the client punishes the advisor for the attack, the client may be considered to be acting as the ego in alliance with the superego, represented by organizational authority.

28. For an analysis of ways in which workers commonly employ these defenses against organizational anxiety, see Jaques (1955), who applies Klein's (1975) theorizing about individuals to organizational situations.

29. For a discussion of ambivalence toward scapegoats, see Taylor and Rey (1953) and Toker (1972). In this connection it is useful to note Freud's (1913; rep. 1950) observation that a person who has not violated any taboo may be taboo anyway if the person is in a state that arouses forbidden desires in others and awakens a conflict of ambivalence. Thus an advisor may arouse envy for the possibility of independence, including the freedom to know and utter secrets and the freedom to criticize. Any affection or admiration accompanying this envy may be uncertainly balanced by unconscious hostility.

8

The Advisor as Reluctant Scapegoat

Interviewer: What other occupations [besides planning] have you seriously considered?

Planner 1: Theatrical technology . . . like working behind the scenes at the theater.

Interviewer: How would you summarize what you consider to be the major principles of professional planning ethics in terms of your work?

Planner 2: A planner is a stage designer. With the preparation of the set or stage, he will certainly screw up the production of the play, but he cannot guarantee a success. He can set the stage, but it is really the actors who guarantee the success of the production. It is the decisions which people make for themselves which are most important. Not only is it not in the purview of planners to make these decisions; it is also not in the purview of the government. Not only are planners not dealing with the most important things; they should not have an inflated view of their importance. . . . I think planners are social workers who don't like dealing with people directly. (Planners in local agencies)

Advising is an important part of bureaucratic problem solving. People ask one another how to define problems, where to find relevant information, how problems arose, who has stakes in situations, where to look for possible solutions, why past efforts went wrong, whether proposals are feasible, and what staff members should do now. But even though collaborative consultation may lead to creative actions, workers with problems often avoid or obstruct others' assistance, for reasons just described. As a result, people asked for advice may respond cautiously. This chapter examines two questions: How do advisors react to clients' ambivalent expectations? And, how may advisors themselves hinder problem solving?

Planners illustrate the dilemmas of advising. Within organizations, planners serve an advisory function to administrators who are responsible for adopting and implementing plans. In the government, planning departments advise elected officials, who may adopt and implement plans through operating agencies. These planners' clients may be staff members of other public agencies, such as the administrators and subordinate staff described in this book, as well as collective public clients such as community groups and their representatives.[1] Although in these cases planners focus on physical and economic development, their concerns exemplify advising dilemmas in general.

This exploration begins with the statements of the two planners quoted above. The first planner works for a transportation agency. Although his job title designates him an analyst because he does research and writes reports on day-to-day problems, he hesitates to call himself a planner. He is essentially an in-house consultant for special problems at the agency, and his assignments require him to discuss problems and proposals with others in the agency, staff in other organizations, and members of the public. He estimates that his time is evenly split between technical analysis or writing and personal consultation.

The second planner works for a state planning agency, where he manages the agency's data base, which is used by local governments in their planning efforts. Thus he is a consultant to city and county planning departments, as well as other public agencies. He emphasizes that, although his agency has little formal leverage over localities, his control of vital information provides it with some bargaining power. He supervises a staff of 15 and estimates that his time is divided about equally between technical work and programming, staff supervision, and consultation with agencies requesting information.

THE PLANNER AS STAGE DESIGNER

These planners describe themselves as stage designers, working backstage to facilitate others' performances. What might this dramatic imagery tell about responses to requests for advice?

In the earlier cases of Jones and Smith, the imagery seems literal. Both are conscious of performing, or putting on a show. At the same time, the drama not only includes consciously chosen actions, but also plots, roles, and actions of which the "producer" is unaware.

Still, even if Jones and Smith are not conscious of everything they do, their use of a dramatic metaphor to formulate their actions is extraordinarily insightful. This metaphor expresses authorship for actions. Psychologically, the metaphor claims responsibility for activities even if the actor is not conscious of all actions performed.

The two planners quoted here are also exceptional in their references to a dramatic role. Interestingly, in contrast with Jones and Smith, who regard themselves as center-stage actors, these two see themselves backstage. They say that they do not act in dramas they help to produce. As will be seen, this assertion of a backstage position is an effort to disclaim responsibility for the consequences of actions. This metaphorical description of the role and discussions of stage design particularly provide hints about planners' perceptions and unconscious intentions that may underlie or influence conscious advice-giving roles.

With both planners, the image of the stage designer sets themselves off from the actors. The first planner speaks of "theatrical technology," an undeniably important ingredient in any theatrical production but one which sounds quite different from the artistic activities. In addition, the planner speaks of "working behind the scenes." This phrase suggests several interpretations. One is the physical separation from the actors. At the same time, there is also a suggestion of power, as of the "power behind the throne." There is even an implication of deviousness—people "working behind the scenes" and not "up front" in their actions. Finally, "working behind the scenes" implies invisibility. Thus this seemingly simple comment combines a number of themes and apparent motives. There is a wish for power over others' actions linked to a wish to avoid direct contact with them or responsibility for the outcomes of their actions, along with a desire for invisibility.[2]

The second planner suggests some reasons for these motives. He, too, depicts the planner as a stage designer. However, if the planner acts alone, in preparing the stage, he is likely to "screw up," or spoil, the production. In contrast, the actors in front, on stage, can "guarantee the success of the production." This is not simply a division of role and location, but also a difference in competence and responsibility.[3] This planner describes the anticipation of ineptness, which is accompanied by unnamed feelings of doubt and shame. He assuages his discomfort by espousing democracy and humility: the actors should make important decisions for themselves, and planners "should not have an inflated view of their importance." Nevertheless, the final sentence expresses the wish to come out from behind the scenes and work "with people directly" in actions that would "guarantee the success of the production."

Both planners express the wish to have power over the success of action on the stage. But the second planner doubts his ability to contribute to such success, and voices a feeling of shame in response to this doubt. In this light, then, the separation of the planner/stage designer from the actors is an understandable defensive position. Suggestions of "behind the scenes" deviousness may be interpreted as caution—evasion of direct responsibility for the outcome is self-preservation. If this desire to escape culpability is taken to an extreme, it can lead to a wish for invisibility.[4]

These metaphorical statements offer hints about planners' anticipations of the advising relationship: they may wish to have power over the implementation of solutions, but may expect others will get any credit, while planners will be blamed for failures. Significantly, these assumptions mirror the conditions of scapegoating. The scapegoat is placed in the center of a stage designed by others. It is assigned great responsibility for others' problems and is powerless to refuse, or even modify, the role. If the play succeeds, others receive the credit; the scapegoat gets only postmortem appreciation.

Aspects of this ritual scapegoating are repeated in advising. Clients may demand a lot of work from advisors, only to ignore the work and dismiss the advisors. Clients may blame consultants for not solving problems, or even, occasionally, for causing them. Advising has none of the overt violence of ritual scapegoating; clients deliver blows only to their consultants' egos. However, when clients unconsciously enact portions of the scapegoating ritual described earlier, consultants may find it hard to work competently and difficult to be effective. Thus a planner's wish to be a stage designer is understandable. In reaction to the risk of being responsible without power, efforts to be powerful without responsibility provide reasonable protection. This chapter identifies specific expressions of this defense when people are asked for advice.[5]

PLANNING AGAINST BEING SCAPEGOATED

If people asking for advice unconsciously attempt to scapegoat consultants, advisors may respond in several ways. Those who realize that the behavior is self-defeating may attempt to explicate and discuss its difficulties with their clients. If the consultants fail to make these actions discussable, they will find themselves forced to deal with the actions covertly and perhaps unsuccessfully. Others may not see the traps at all and may be drawn toward them. They may unconsciously recognize the invitation to act as scapegoat and may unconsciously select defenses against it. However, because these calculations are not overt, they cannot be tested for realism and may only lead to further problems.

Unfortunately, the conditions under which bureaucratic workers are most likely to attempt to scapegoat advisors also make it especially difficult for advisors to recognize such an unconscious agenda or make it overt. Thus Argyris and Schön (1974, 1978; Argyris, 1982), studying planners and other consultants, find nearly universal defensiveness and resistance to discussing clients' tacit, even unrealistic and self-defeating, assumptions. Advisors commonly react to their clients' covert intentions simply by concealing their own. Lipsky (1980) observes that "street-level bureaucrats," including planners, are especially likely to act defensively and secretively when they face excessive demands and have limited resources to offer.[6]

Practitioners' educational backgrounds also affect their ability to perceive and respond constructively to clients' unconscious intentions. Advisors with training in such fields as psychology, the social sciences, social work, nursing, counseling, teaching, or some administration may be most comfortable with and sensitive to both their clients' and their own feelings and may be most insightful in interpreting the covert content of relationships. In contrast, advisors with more technical training, such as engineering, economics, mathematics, the natural sciences, medicine, and some architecture, are less likely to be comfortable with or sensitive to feelings and, consequently, may be more susceptible to unconscious psychological dynamics. Both formal training and personality orientations contribute to these differences (Holland, 1973, 1978). Thus, although unconscious scapegoating intentions are insidiously seductive, consultants from technical fields may be most vulnerable. Interestingly, the first planner quoted earlier is an architect, while the second graduated in sociology, and both reveal concerns about scapegoating.[7]

With these variations, an advisor who senses a client's intentions to scapegoat may react defensively by attempting to reverse the unequal distribution of responsibility and authority and to obscure the advisor's own responsibility. In doing this, a consultant may attempt to alter the client's views of the consultant's role or perceptions of the problem-solving process. The following examples from planning have counterparts in other fields.

Ritualism

Ritualism, a defensive tactic discussed in Chapter 6, involves goal displacement. Instead of treating bureaucratic regulations or technical standards as a means of solving problems, people emphasize the enforcement of regulations or standards as ends in themselves (Merton, 1940). If asked to rationalize their actions, workers would have to argue that what they are doing is magical, like a ritual—if they perform the prescribed actions often enough, they should have some beneficial effect. At least unconsciously, they recognize that the mere repetition of innocuous actions is likely to avoid trouble.

Ritualism requires an advisor to redefine him- or herself from an active thinker to a passive interpreter of rules. A ritualistic planner claims the role of stage designer and insists that he or she is only following someone else's script. In this role an advisor may portray problem solving as simply the application of standards derived from professional norms. For example, planners may present proposals and justify them as unalterable because they are based on technical rationality, economic efficiency, or design aesthetics. A planner may insist on the unassailability of any "optimal" solution derived from rigorous quantitative calculations. Similarly, a planner reviewing development proposals may render authoritative judgments based on expert interpretations of standards for height, bulk, siting, parking, cost, and the like. In all these instances, planners act only as the agents of rational principles, models, and methodologies, which themselves exert power.

Alternatively, planners may emphasize their role as enforcers of bureaucratic regulations. For example, in either supporting their own recommendations or reviewing others' proposals, planners may authoritatively interpret regulations in zoning, health care, transportation, or whatever the relevant field in insisting that only a single position is acceptable. In this way planners may exercise all the power legally enacted into the regulations without personal responsibility for decisions or their consequences. Schön's (1982) case study of the planner who reviews development applications, summarized in Chapter 6, illustrates this approach. That planner recognizes situations as legally ambiguous, but he pretends to others that he only deduces interpretations from inclusive regulations.

Planners often have discretion in interpreting standards and regulations, as well as in deciding whether to invoke standards and regulations at all. However, these norms and rules offer power in making recommendations while cloaking responsibility in impersonal sources. Without personal responsibility, planners may avoid being scapegoated.

Task Redefinition

Planners may also defend themselves by presenting the planning process as something other than the planner solving a client's problem. One approach is to displace emphasis from the outcome of the planner's efforts to the work process itself. While an advisor may bear responsibility for the outcome, the burden of responsibility for the process may be laid upon the client. To do this, a planner may emphasize the structure of the consulting relationship, the client's cooperation, or the client's demeanor. For example, planners may insist upon the completion of a great amount of paperwork before giving advice. Planners who work with community groups may focus on residents' style or language. For example, planners may say it is difficult to consider the points of view of local residents when they fixate on parochial neighborhood concerns rather than the interests of the metropolitan public. Residents' "emotional" language is not useful in developing a rational analysis of problems. If the planner never has to submit a recommendation, then there is nothing for which he or she can be blamed. If responsibility for initiating or solidifying the advising relationship rests upon would-be clients, difficulties that arise may be attributed to client improprieties.

A related tactic in shifting responsibility is to redefine the division of labor in the work process. Advisors may emphasize that they are less powerful than clients imagine and that a great deal depends on the client's investment. This valid observation may be a helpful response to a client's extreme dependency, but it may also be simply setting up the client to take the blame for later problems. For example, if a community group requests assistance in developing a primary care center, the planner may see several obstacles to a workable proposal—residents disagree about the purpose of a center, need for a center is not evident, funding is doubtful, and recruitment of good management would be difficult. Instead of confronting residents with these concerns and risking disappointment and criticism, the planner may emphasize participatory planning and self-help as a way of leaving planning responsibility with the citizen group. Once they have prepared an initial proposal, the planner will respond by criticizing their suggestions.[8]

Alternatively, advisors may attempt to redefine expectations about outcomes to match what they are certain they can accomplish. For example, a consultant working with a builder may tell the client that as zoning regulations are so difficult to modify, any variance obtained should be accepted as a major accomplishment. Planners working with community groups may emphasize the complexity of the planning process and the power of other groups as a means of preparing community members for limited gains.[9] Planners working on highly politicized issues may emphasize that the political issues are someone else's responsibility and beyond technicians' control. Thus when planners refer disdainfully to "politics," they both express disagreement with a specific way of solving problems and proclaim limits to what should be expected of them. Another way in which advisors may redefine outcomes is to convert them into different currency. For example, some planners may emphasize their own good faith and the solidarity of their relationship with clients, regardless of whether physical or economic development ensues.[10] Implicitly the redefinition of expectable outcomes is a voluntary surrender of competence by the planner, but this loss may be acceptable if it avoids disappointments for which the planner may be blamed.

In each of these instances, a planner may covertly exercise power. He or she may attempt to influence the choice of a solution for a problem or may shape the client's expectations of the planner and the problem-solving process (Forester, 1982). In this way planners can shift responsibility for problems onto clients. These tactics may help advisors avoid confrontation with clients, blame for problems, and frustration, reprimand, or dismissal. These planners respond to risks of being scapegoated by setting up their clients, instead, to face blame or harm.

INADVERTENT ATTRACTIONS TO THE SCAPEGOAT ROLE

Reasons to avoid being scapegoated are self-evident. Nevertheless, advisors may be attracted to certain behavior that "nominates" them for scapegoating. Some of these attractions stem from the imperatives of problem solving; others may develop from the advising relationship.

Tendencies of Problem Solving

Whenever a problem is tied up with "secrets" or covert assumptions, an advisor confronts a double bind. Clients may overtly ask the advisor to do anything necessary to solve the problem, but tacitly ask him or her not to disturb their position or disrupt the organization. Following the latter request is likely to lead to the advisor's failure, but if investigating a problem leads to exposing "secrets" and covert political conflicts, the advisor is likely to be blamed for them.

Conventional problem-solving expectations, which encourage advisors to say things others do not, fosters the likelihood that the advisor identifies conflictual "secrets." A planner, for example, is asked to advise because of expertise which others lack. In particular, planners' ideal of comprehensive rationality encourages them to collect, analyze, and report as much information as possible. At the same time, planners' emphasis on technical, rather than political, information may have either of two effects. Insofar as planners restrict themselves to technical information, they may avoid discussing controversial political issues that make them the center of conflict and link them with blame. However, many planners attempt to avoid political issues even while they only poorly understand the political meanings of the problems on which they work (Baum, 1983a). Thus they do not know what to avoid, and they may unintentionally proclaim the importance of concerns they do not recognize as the focus of political contention. Their very contempt for "politics" may lead them to ignore suggestions to withdraw.

For example, a team of planners reviewing alternative subway station sites may focus on commuting patterns and locational potentials for economic development, while overlooking the mayor's interests in allocating stations to political constituencies. Certainly the planners' emphasis is valid. However, unless they recognize the likely consequences of their work and are prepared to anticipate and respond to them, at best, they may be disregarded, or, at least as likely, they may be blamed for any disappointments in the designation of stations.[11] In this way a single-minded focus on a client's ostensible problem may lead to being scapegoated for failure to solve a political problem.

Responses to the Advising Relationship

In addition, just as unconscious meanings of the advising relationship may lead clients to scapegoat consultants, these meanings may lead a consultant to act in ways that inadvertently lead to scapegoating. Such responses correspond to clients' transferences in the advice-seeking relationship and may be considered consultants' "countertransferences." The consultants either accept clients' erroneous assumptions about the relationship or add their own unrealistic assumptions to it.[12] These countertransferences involve assumptions that appear to defend advisors against the risks of problem solving and offer control over the process. However, as a result, advisors act in irrational ways that not only conflict with comprehensive rationality but also make them likely sacrifices after ensuing failures.

For example, a consultant may wish to ease a client's shame about having a problem. As a result, the consultant may limit the investigation of the problem. The advisor may collude with the client in taking the view that few alternatives are feasible. For instance, consultants advising a city government that has made several disastrous, nearly scandalous, ventures in solid waste disposal may focus on assuaging the mayor's public embarrassment and avoid investigating the reasons for the succession of failures. Consequently, the consultants may recommend a conventional facility which is as likely to fail as the others because it does not deal with the reasons for the previous failures. If the proposal is accepted, the city will waste still more money, and city leaders are certain to criticize the consultant for the failure.

When a planner invests considerable time in developing proposals that should succeed, he or she may lobby for public recognition for the proposals. Further, the planner may believe that, without his or her control, public agencies, development corporations, or community groups are unlikely to implement the proposal effectively. Although this self-confidence and these perceptions may be realistic, they may also be nurtured by clients' fantasies of extreme dependency on the planner. The planner may tacitly accept their fantasies that he or she alone has the competence to solve problems. These expectations can only be disappointed. For one thing, no one can satisfy such high hopes. In addition, a client who operates on such faith in the power of a consultant is unlikely to provide the effort and information really necessary for developing workable proposals. Thus a planner who pushes for so much responsibility may simply be moving into a position to be criticized and dismissed for what must be failures.

An advisor who works for a long time with a client may want to become accepted as an insider. These wishes may be encouraged by clients' involvement in conflicts with others, so that the advisor's boundary position makes him or her look like a possible adversary. The advisor may accept the client's tacit invitation to serve as a leader in fighting—or, if necessary, fleeing from—other political actors. In the process, the advisor may transfer his or her own unconscious assumptions about authority onto the client's relations with others.

For example, even though Simon, described previously, portrays himself as the object of a seduction and setup, it is possible that he unconsciously participated in a mutual seduction, in which he offered his knowledge in return for insider status and intimacy. He may have unconsciously wished to set himself up to challenge the boss because of hidden interests in unseating fatherlike people in authority. Although the offer of the earlier consultant's letter placed him in a bind, he may have requested the letter partly to secure an unconscious alliance with the assistant director against this authority figure. In these ways, he might have been attempting to rescue himself from an anxiety-producing boundary position by redefining the situation in unconsciously familiar and more satisfying ways, perhaps a wished-for alliance with mother against father.

In another example, a neighborhood group may persuade a community planner to criticize local banks and realtors as hindrances to minority homeownership. The planner may be moved by a realistic perception that these institutions are obstacles. However, the planner may also unconsciously think of this conflict primarily as an opportunity to challenge people with authority and get pleasure simply from making an alliance with subordinates against their superiors. Although this unconscious interest may help sustain action in a protracted conflict, it may create problems insofar as it distracts from a realistic analysis of political relations with powerful institutions and precludes formulating a workable strategy. Certainly planners have ethical responsibilities to challenge discriminatory practices, such as redlining. However, unless planners realistically assess these practices and their own interests in attacking community institutions, they may be trapped into blame for problems. For example, the planner may not recognize the risk of community members abandoning him or her at the first sign of counterattack from the banks, realtors, or elected officials. Those institutions may then identify the planner as the assailant, while community members now treat the planner as a troublesome irritant. Both sides may focus on eliminating the planner, rather than considering problems of homeownership.[13]

A planner may continue an ill-advised attack on powerful institutions or administrators because the assault offers an illusion of power. Clients may persuade the planner that publicizing certain political "secrets"—such as the mayor's tacit support for lenders who restrict minority homeownership—will automatically empower the planner to make major changes. Although this expectation may have some realistic basis, it may also be nurtured by clients' fantasies about consultants' power and shared fantasies about the power of "secrets." Planners may imagine that broadcasting clandestine political affairs will unseat community institutions in the same way that children fantasize that spreading the knowledge of parents' sexuality disempowers them. However, a planner who focuses on publicizing covert political activities while avoiding realistic strategies risks becoming a welcome community sacrifice.

The Paradoxical Role of the Scapegoat

Bureaucrats, officials, and community groups engage advisors to provide powerful assistance in solving problems. However, organizational psychodynamics, as well as the dynamics of advising relationships, encourage these clients to weaken their advisors as well. An advisor who responds to these unconscious wishes may find him- or herself drawn toward an apparently self-destructive role. At the same time, although the scapegoat role may defeat an advisor's conscious intentions, it offers two unconscious attractions that may outweigh its obvious risks. The first concerns an advisor's wish to solve problems. The second concerns a wish to avoid the dangers of intimacy in collaboration that may be necessary for solving problems.

If scapegoating conflicts with interests in instrumental solutions for problems, the process of preparing an advisor to be scapegoated may be unconscious. During consultation, one unconscious assumption most consistent with conscious intentions may dominate. A client may unconsciously regard the potential scapegoat as powerful and efficacious in a special sense; the sacrifice of the scapegoat seems to solve problems as if by magic. The client may not be aware of making this assumption, and consequently cannot communicate it explicitly to an advisor, but nevertheless may tacitly lead the advisor to believe that becoming a scapegoat brings magical power. An advisor interested in solving problems may unconsciously move toward this role, confusing magic and instrumentality.

Thus an advisor may find him- or herself unaccountably persuaded by arguments that challenging others, especially those in authority, will provide the power needed for solving problems. Despite doubts about this possibility, belief in it may be reinforced by an unconscious communication that the very self-destruction in this challenge holds the key to an expiatory solution to the problems. Some of the role's attractions may be suggested by a description of Indian Brahmans:

> In Travancore, when a Rajah is near his end, they seek out a holy Brahman, who consents to take upon himself the sins of the dying man in consideration of the sum of ten thousand rupees. Thus prepared to immolate himself on the altar of duty, the saint is introduced into the chamber of death, and closely embraces the dying Rajah, saying to him, "O King, I undertake to bear all your sins and diseases. May your Highness live long and reign happily." Having thus taken to himself the sins of the sufferer, he is sent away from the country and never more allowed to return. (Frazer, 1940, p. 543)

Unlike the scapegoats described earlier, and more like contemporary consultants, this scapegoat is not killed, but is only banished. The Brahman agrees to undergo banishment not only for monetary compensation, but for the prestige of absolving the king's sins and attempting to extend the king's life. These thoughts hint at consultants' unconscious considerations. The holy man takes on the king's sins in his embrace but must then be removed from the community. Crucially, as consultants may unconsciously assume, this expulsion succeeds in removing the sins as well.

The magical expiatory power of the scapegoat role balances and compensates for the absence of the instrumental power an advisor may want. This reward makes the role seem more attractive and less self-destructive. These actions may, in fact, bring some overt power, but it is limited by the advisor's marginal position in the organization or community. Someone who utters organizational or political "secrets" has real, even if limited, power to disrupt. For example, a planner who identifies the role of local banks in preventing black homeownership threatens the banks' business. Even though the planner may be powerless to increase homeownership, he or she may have the power to disturb normal activities of the banks, associated real estate interests, or local officials. The planner becomes a center of attention and must be responded to. The same is true of the advisor who reveals bureaucratic "secrets." This person cannot be simply ignored but must be confronted, and the "secrets" must be suppressed once more.[14]

Still, the negative power is self-limiting. Once the advisor expresses ineffable points-of-view, others withdraw and exclude the advisor from further collaboration. Thus the consultant loses any possible responsibility for solving problems. Although this development conclusively deprives a consultant of creative problem-solving power, it may make the role unconsciously more attractive. Facing the trap of responsibility without power, the consultant acquires both the power to disrupt and magical expiatory power while shedding any responsibility for the outcome.

In addition to being freed from culpability, the consultant avoids the emotional risks of developing powerful collaborative arrangements with others. The separation of the scapegoat may rescue an advisor from the anxiety associated with the uncertainties and ambiguities of day-to-day work with a client. The second planner quoted above says that "planners are social workers who don't like dealing with people directly." In this way, paradoxically, embracing the scapegoat role may defend an advisor against fears about being drawn into the role. For now there is no longer any reason to risk entering into closeness with clients to share information, work out relationships, discover and make mistakes, or move toward agreements. An advisor need not worry about taking responsibility for others' concerns or interests, defending others or their positions, being deceived or exploited by others, being misunderstood or embarrassed by them, or simply taking firm positions that could be wrong. It is easier to conclude that, somehow, scapegoating is inevitable, that the client does not really want advice but just wants a whipping boy.

Avoiding the risks of Type IV power relationships, an advisor may retreat past the dangers of Type III conflict to a Type II position, in which personal control over "good" information and insight may bring secure satisfaction. Still, although such a retreat may protect an advisor from the risks of everyday advice giving, it also deprives the advisor of the possibilities of being competent. For this deep loss, a consultant may decide to punish the client, and here too, the scapegoat role "works." For when an advisor loses competence in embracing the scapegoat role, he or she withdraws competence from the client as well. Thus an advisor may take revenge on a client by becoming as weak and ignorant as the client hinted all along.[15]

WHAT DOES IT MEAN TO CALL THIS SCAPEGOATING?

The extreme connotations of "scapegoating" should not mislead one to conclude that the advising difficulties described here are themselves extreme, either in violence or in frequency. To the contrary, they are commonplace. Nearly everyone who has ever been asked to advise has worried about humiliation, ineffectiveness, criticism, and dismissal. Describing the dynamics of advising as intentions to scapegoat and efforts to resist it identifies dominant themes in consulting and offers some explanations for consulting conflicts.

People's interests in satisfying psychological contracts create expectations about relationships in work organizations. Bureaucratic psychological structures frustrate and subvert many of these interests, turning them, instead, into wishes to escape, create games, or avenge disappointments on others. Bureaucratic workers confront advisors at moments many workers find particularly painful. The subordinates have problems they cannot solve alone. Others may blame them for the problems, or they may blame themselves. They may feel that their ability to solve the problems is a trial for organizational acceptance, as well as a test of personal competence. Both real supervision and internal insecurity may superimpose considerable anxiety on problems which are complicated. Advisors, in turn, may "catch" their clients' anxieties and may share similar feelings. Advisors are expected, and they expect themselves, to solve difficult problems in a short time.

When the combination of personal wishes and organization situations creates anxiety the scapegoat drama is available to both clients and advisors for managing their feelings and thinking about their relationship. The drama represents a common cultural myth people preserve for its power in organizing, containing, and resolving recurrent emotional conflicts.[16] Crucially, this means that withholding information, ignoring recommendations, and fruitlessly attacking bosses or public institutions may not simply be expressions of individual incompetence or random accidents. People working on problems may do many of these things as part of a more or less coherent collaborative unconscious response to systematic organizational conflicts.

The Powerlessness of Scapegoating

Still, it is clear that, while this play responds to some problems, it creates others. Obviously, when people become more concerned about scapegoating than about analyzing problems, they are unlikely to work together. For both client and advisor collaborative power remains a mystery. Each may be interested in acting powerfully, but more likely against one another than together. Neither advances past the confusion about power expressed earlier in Chapter 2. Participation in any of the elements of scapegoating only confirms them in their assumptions that becoming powerful means defeating others, that others have power and are likely to exercise it punitively, and that their only chance to exercise power themselves is through stealth. Either in the embrace of combat or in retreat from it, neither learns how to collaborate powerfully, and both fail at solving many organizational problems.

At the same time, workers differ in both extent and content of anxiety. A worker who feels anxious may confront a consultant who does not, and vice versa. When client, consultant, or both feel anxious about their work, both, one, or neither may enter into a scapegoat drama to organize their anxiety. Even if both begin to interpret their relationship in terms of scapegoating, they may have different versions in mind, even though the basic themes are constant.

The concluding chapter focuses on understanding why some workers can collaborate powerfully and solve problems, and learning how to increase the chances of this. The discussion examines changes in bureaucracies, and members' assumptions about them that may facilitate organizational problem solving.

NOTES

1. Private planning consultants may work either for private individuals or for public agencies.

2. Certainly, the validity of this interpretation should be examined in discussion with the planner who made these statements. One type of evidence in support of this interpretation is the consistency between these comments about other occupations he had considered and other images of planning presented elsewhere in the interview.

3. This differentiation between the competent and the irresponsible recalls Menzies' (1975) observations of nurses' stratagems in assigning competence to their superiors as a means of avoiding blame for feared failure. See Chapters 3 and 4.

4. On the other hand, it is possible that it is the stage designer's location "behind the scenes"—in the "bowels" of the theater—that gives rise to feelings of shame. Further, the stage designer may believe that noxious actions from the stage contaminate the designer's efforts and make success doubtful.

5. May (1972) presents an illuminating case study of such a reparative effort. He describes a man who gradually constructs a fantasy realm in which he exercises growing power over others without responsibility for the consequences of his actions. May discovered that the man's early life had been just the opposite. He had been saddled with great responsibility for others and had lacked the power to choose, modify, or reject any specific responsibilities. The adult fantasy world provided a symbolic remedy for a barely tolerable earlier situation.

6. Lipsky's portrayal of bureaucratic workers resembles that of this book, with two differences. First, he concentrates on direct service workers' relations with public clients, rather than bureaucratic clients or bureaucratic superiors. Second, his analysis is primarily sociological, rather than psychological. The similarities of his findings tend to confirm the validity of the descriptions here, while his sociological perspective introduces additional, complementary explanations.

7. One study of city planners' personality orientations (Baum, 1983a) suggests that these practitioners include both "social" and "technical" subgroups.

8. See Checkoway (1981) for related examples.

9. See Needleman and Needleman (1974) for other examples.

10. See Baum (1983a) for other examples.

11. For examples of planners who intentionally and strategically become involved in political controversies, see Clavel (1986), Krumholz (1978, 1982), Krumholz and Cogger (1985), and Krumholz, Cogger, and Linner (1975).

12. For a discussion of countertransference in clinical relationships, see Kernberg (1965), Orr (1954), Racker (1968), Reich (1951), and Winnicott (1958). For an analysis of what might be considered countertransference in organizational relations—superordinates' responses to subordinates' transferences—see Hodgson, Levinson, and Zaleznik (1965).

13. For case examples of neighborhood planners' challenges to local political institutions, as well as the risks of planners' being drawn in to fight others' battles, see Needleman and Needleman (1974).

14. This drama may be conceptualized as a conflict between superego and id interests. The advisor revealing "secrets" may be regarded as a representative of repressed material, which bureaucrats and officials recognize and believe that reality requires them to defend against and once more repress.

15. This final act suggests one more meaning to being "behind the scenes," by recalling Freud's (1923, rep. 1962; 1930, rep. 1965) topographic model of the mind. What is onstage may be considered conscious, while what is backstage is unconscious. Conflicts between planners behind the stage and their clients onstage may be interpreted as conflicts in which unconscious organizational or community interests insist upon conscious representation and employ deviousness or treachery when direct expression is denied.

16. Jungians (for example, Jacobi, 1973; Mitroff, 1983) would assign the drama to a "collective unconscious." As such, it is always present deep in an unconscious realm that everyone shares, and periodically emerges into consciousness to organize people's actions and thoughts about their actions.

9

Learning and Problem Solving

Psychoanalysis leans toward . . . values that support the recognition of the depths of the inner world, complexity, ambiguity, conflict, the ubiquity of the demonic and of suffering, the frequent interpenetration of victory and defeat, unremitting questioning of absolutes, and the like. (Roy Schafer, *A New Language for Psychoanalysis*)

Organizations vary. Some are nearly totally bureaucratic, while others contain very few bureaucratic elements. Workers vary as well. They expect different rewards, and they react differently to organizations. Some deftly collaborate with others in solving complicated problems, while others struggle with themselves over whether to collaborate and, if so, how to frame problems safely. There is no certain way to say what proportions of workers fall into each of these two broad categories, although workers' stories suggest that many have conflicts at least some of the time.

Both groups can teach us something about organizational problem solving. Preceding chapters have focused on workers with conflicts because their difficulties help to understand how bureaucracy impedes or complicates problem solving. These workers are exceptionally well-educated; emotionally, they are normal. By analyzing what can go wrong even for them, we can identify what organizational competence requires for workers generally. At the same time, bureaucrats who collaborate successfully demonstrate that conflicts are not inevitable and offer lessons about conditions under which problem solving is possible. In some cases, these workers are exceptional individuals. More often, they use their abilities insightfully to discover, create, and exploit organizational possibilities. Occasionally they work with others to change formal organizational structures. This chapter proposes some guidelines for more effective and satisfying work.

The psychoanalytic perspective on organizations offers workers possibilities for competence by revealing a more accurate, richer organizational map. In shedding light on unconscious meanings of bureaucratic relationships, this perspective helps workers identify aspects of organizations to which they normally respond unconsciously. By opening these meanings to examination, this orientation helps workers recognize more of what really influences them, including their thoughts and feelings about colleagues. They may deliberately endorse some of their normally tacit assumptions and reject others as self-defeating. They may then more realistically assess and deploy resources available to them.

One of the fundamental points of this book is that this psychoanalytic perspective is rational. It simply extends the problem-solving ideal that most practitioners already recognize and espouse. This perspective shares the norm that people should select actions on the basis of their strategic contribution to realistically chosen goals, but importantly, the psychoanalytic approach recognizes unconscious data in addition to the material on which most interpretations of the rational model focus. Thus this perspective offers workers a basis for truly comprehensive information collection and analysis. Then workers can do just what textbook decision making models promise—rationalize choice and promote control through the examination of all relevant data.

Preceding chapters show that bureaucratic workers recognize that they do not always act effectively, though they want to do competent work. Most workers at least occasionally are ambivalent about what they do. This is the central point about psychological conflict—bureaucrats may have difficulty solving problems not because they attempt too little, but because they attempt a great deal, some of it in opposition to the rest of it. They may fail not because they do not try hard enough, but because they try with great commitment, only in several opposing directions at once. The psychoanalytic orientation can shed light on these conflicting aims and show workers how to promote their conscious goals with less interference.

However, when people feel ambivalent about their actions, they also feel ambivalent about thinking about those actions. Hence, despite the appeal of new, psychoanalytic knowledge, bureaucrats may want to avoid examining their conflicts because doing so would require them to look more closely at situations that make them anxious. The next section of the chapter analyzes reasons why bureaucratic workers might resist knowing more about psychological aspects of organizations. The following section presents guidelines for intervention which workers could follow to increase their effectiveness and satisfaction. The book concludes with an examination of conceptual, methodological, and ethical issues raised by a psychoanalytic orientation to organizational analysis and intervention.

RESISTANCE TO PSYCHOANALYTIC KNOWLEDGE

Alleviation of the difficulties portrayed in preceding chapters requires more than the adoption of new managerial, organizational, or planning techniques. Bureaucratic workers must be willing to think differently about their experiences; they must be willing to recognize that they have unconscious thoughts and feelings about other people and that these thoughts and feelings affect their competence. And yet many workers may find this new knowledge unsettling, as the following case of a hypothetical city planner illustrates.

"Mr. Turner" plans recreational facilities for a municipal planning department. As many practitioners, he attempts to incorporate a variation of a textbook model of rational planning into his work. He follows a logic of defining a problem, setting goals for its solution, identifying alternative strategies to meet the goals, and selecting an optimal strategy according to consistent criteria. Although he acknowledges the impossibility of complete information collection, he attempts to collect a great deal of information, consulting many published sources and conferring with other agency staff and citizen groups. When he meets with community groups, he gets involved in conflicts over the types of facilities that should be included in each neighborhood and where particular facilities should be located. Although he recognizes the value of citizen participation, he insists that discussions at public meetings be balanced by systematic surveys of residents and rigorous analysis of locational alternatives. If local politicians attempt to influence decisions, he deliberately ignores their positions, just as he resists getting involved in bureaucratic politics. In the end, he tends to rely mainly on his own interpretations of published recreation standards in making recommendations.

As Turner struggles with difficult choices, the psychoanalytic point of view raises questions whether some things—including his actions—are not still more complicated than they seem. Perhaps his reactions to politicians, community groups, and bureaucratic officials include negative reactions to his parents or siblings. Perhaps he emphasizes objective standards and analysis over residents' political demands because he feels anxious about "giving in" to people. Perhaps his strong assertion of a rational planning process is partly a reaction to feelings of weakness before "politicians." Even though he proposes facilities that community groups often praise, he might find that anxiety keeps him from putting forth some really innovative ideas and from challenging conservative opposition from elected officials. Despite his successes, he might find such discoveries humiliating. If he already feels ashamed before bureaucratic authority, understandably, he might resist knowing more.[1]

Turner's hesitation about learning may be illustrated in connection with his hostility toward "politics" and "politicians," something relatively common among planners and other bureaucrats. It is possible that what he sees in politicians and dislikes about them represents some of his own motives which he disapproves of and has unconsciously denied and attributed to others. In particular, he might find that the "irrationality" of the politicians whom he disdains or fears is, in the end, a mixture of his own unconscious wishes. Thus he might be forced to conclude that some of the opposition to his rational planning is within himself, not in other, obviously "unreasonable," people.

Such a finding threatens his feelings of competence, autonomy, and control. Working with community groups, politicians, private developers, and agency staff challenges him intellectually and politically to manage the planning process so as to get consensus about problems and their solutions. The textbook model of the rational problem-solving process provides guidance in identifying relevant information and points when decisions should be made. It brings a modicum of control to an often turbulent setting. Considering the unconscious and conflicting aspects of this situation threatens Turner's belief in his control. His dominance over the planning process may now seem illusory. For example, while he has thought of himself as a master of rationality, he might have been motivated less by conscious rational planning goals than by unconscious anxiety about avoiding conflict. In addition, he may unconsciously wish to act "irrationally"—self-interestedly, exploitatively, duplicitously—and his "defeats" by politicians might be motivated by unconscious desires to let them act out his "improper" wishes. Any such discoveries challenge his sense of competence, for even some of his evident successes may seem accidental. They are consistent with his conscious aims, and yet unconsciously he has also worked against these aims, so that he cannot be sure how much of the success he can claim credit for.

On the one hand, the psychoanalytic view is sympathetic to Turner's conscious intention to plan rationally, seeing him as an occasional unwitting victim of the deviousness of his unconscious desires. However, this insight does not help him with members of the public, who have little interest in what may be conscious or unconscious. They hold him responsible for all his action, including failures. The psychoanalytic perspective supports the public view—it, too, holds him responsible for his unconscious, irrational aims and their consequences. Turner may conclude that psychoanalytic knowledge brings him new responsibility for failures he has blamed on others and deprives him of credit for successes he has claimed for himself.

Moreover, these discoveries undercut Turner's usual public arguments for autonomy on the basis of planners' professional goals and ethics. The psychoanalytic view suggests that he, as other practitioners, often does not single-mindedly pursue a model of rational decision making, but is motivated by a combination of rational and occasionally quite irrational interests. In addition, he may not simply be motivated by an overriding concern for a public interest, but may seek to satisfy many personal, private interests as well.

Thus psychoanalytic exploration may unsettle him by making him aware of two domains which he has attempted to ignore: the external organizational and political environment and the internal unconscious realm. Both worlds exist and affect his efforts to solve problems. He is responsible for more than he suspected or wanted. Hence if he does not have a satisfying psychological contract, he cannot blame only the organization, but must examine his own assumptions as well. Although these insights may help him take more conscious control of his actions, they immediately challenge his self-esteem and may discourage him from learning more. Nevertheless, the lesson of psychoanalysis is that knowledge is the prelude to change.

GUIDELINES FOR INTERVENTION

Bureaucratic workers can learn more about psychological obstacles to solving problems and by using this knowledge can change themselves or their organizations to improve their performance. Commonly, people expect that organizations permit work that calls on competence, entails responsibility, carries significant authority, and brings recognition. Personal actions should have meaningful consequences. Regarding peer relations, some workers may want considerable intimacy, whereas others may want only cordiality and reciprocity. However, relations should be meaningful, more or less open, and relatively free of duplicity. Above all, organizations should permit workers to collaborate powerfully in solving problems.

A Framework

A framework for intervention requires clarity about goals, strategies, and tactics. What situations are to be changed? What are the elements of those situations where modifications may change the entire situation? And what actions may modify those elements? Our understanding of bureaucratic psychological dynamics is still limited. Thus interventions should be treated as experiments, from which we may learn more.

The goals of intervention should be to eliminate obstacles to problem solving in organizational psychological structure, resultant predicaments and responses to them, or problem solving approaches. These obstacles are targets for change. Because they represent successive stages in the development of impediments to solving problems, the earlier an intervention is applied, the more likely it is to influence workers' performance.

Each situation to be changed may be regarded as a combination of actions or influences in four domains. The first involves individual thinking, feeling, and action. The second involves interpersonal relationships. The third involves the organizational social structure, either formal structures or consistent, organization-wide patterns of interaction. The fourth domain involves societal structures and cultural norms that influence the structure of social institutions, patterns of personal interaction, or individual thinking and feeling. For example, Fromm (1955) contends that the psychological dynamics associated with bureaucracy result from the alienated relations of a capitalist society. Horney (1973), although not writing specifically of bureaucracy, argues that bureaucraticlike social relations and psychological dynamics are products of what might be called an advanced industrial society. Argyris and Schön (1978; Argyris, 1982) believe that defensive problem-solving assumptions are an outgrowth of socialization in an industrial society.[2] Changes in any of the four domains represent strategic approaches to modifying any of the obstacles to problem solving. As each of the four succeeding domains encompasses the preceding, strategies further along may bring greater leverage. However, this relationship is not automatic, and effective change may require strategies focused on several domains at once.

Overt intervention into any of these domains is a tactic for change. In addition, diagnosing, or learning about, obstacles to problem solving promotes change. This is true in the general sense that it is necessary to define a problem accurately in order to solve it. It is especially true for the psychological conflicts that hinder problem solving. By their nature, unconscious assumptions, thoughts, feelings, actions, and reactions are invisible. People may get pleasure from them at the same time that they consider them selfish, irrational, or unprofessional. Therefore, they may successfully enjoy acting in certain "unreasonable" ways only if they conceal these actions or thoughts from themselves. Accordingly, if individuals are going to change, they are likely to do so only if they can identify the unconscious thoughts that draw them toward ineffectiveness. Once these "unreasonable" thoughts are exposed, they become less convincing and more difficult to follow. Hence, particularly in the psychological realm, diagnosing or learning about the elements of a situation is difficult to separate from more active intervention to change those elements.

Table 3 summarizes this discussion by showing the relationships among these goals, strategies, and tactics. Three observations are important. First, because the obstacles to problem solving are interrelated, some interventions may simultaneously affect several targets. For example, the creation of task groups might modify an organization's psychological structure, reduce its predicaments, and change the use of consultants.

Second, because obstacle situations consist of four domains, some overt actions may simultaneously involve several strategies. For example, establishing regular formal procedures for using advisors might change organizational policy, interpersonal relations among members, and individual workers' assumptions and feelings about their advisors. Analogously, although diagnostic activities may begin with an individual, small group, or a formal organizational unit, learning may pass across these boundaries.

Third, both intellectual analysis and observation of the results of overt actions

Table 3. Framework for Interventions

Strategies/domains of Intervention	Goals/Targets of Intervention		
	Psychological Structure	Predicaments and Responses	Problem Solving Approaches
	Tactics of Intervention		
	Diagnosis/Learning		Overt Action
Individual			
Interpersonal			
Organizational			
Societal			

enable workers to learn more about their organizations and themselves, with the result that they may change their minds about which target to aim at or which strategy to pursue. For example, a worker who starts out analyzing how formal structures stifle his initiative may end up examining how his own feelings constrain him. Alternatively, administators who send their staff to group relations training may find that interpersonal competence is difficult to improve without changing the formal allocation of responsibility and authority.

Examples of interventions appear in Table 4. They are identified with the strategic domain in which they may begin, although they may extend into other domains. Because societal interventions are complex and poorly understood, they are not discussed here. Interventions are not identified with specific targets because the obstacles to problem solving are closely connected, although discussion of some examples will show their particular connection to one target or another.

Many of these interventions are familiar, and they do not necessarily reflect psychological thinking. Reforms may affect organization functions in several ways simultaneously. The proposed actions may reduce unconscious psychodynamic conflicts even if bureaucratic workers and reformers do not think psychologically. However, one reason that many of these common proposals are never adopted is psychological. In many cases workers have resisted changes which would disrupt their unconscious stakes in prevailing practices. Although proponents may argue for increased efficiency or productivity, implementation may depend on accommodating and compensating workers' unconscious losses. Significantly, while these proposals illustrate the interplay of the three domains, they show clearly that individuals have significant control over their reactions to relationships and structures. The following discussion of some of these proposals illustrates possibilities for intervention, as well as constraints on change.

DIAGNOSIS AND LEARNING

Individual

From the beginning of an affiliation with an employer, workers' expectations and assumptions affect their ability to solve problems by framing conditions for a psychological contract. Workers want to be competent and secure. More likely than not, they want recognition; probably they want power of some kind. Some want more intimacy, candor, appreciation, or discretion than others. In any organization some of these expectations are more likely to be satisfied than others because they may be more realistic, more specific, or simply better matched to the setting. Workers' demands of an organization and their reactions to their satisfactions and disappointments mold an organization's psychological structure.

Their unconscious preferences and thoughts contribute to organizational predicaments. Past experiences with authority affect workers' perceptions of and responses to ambiguous bureaucratic responsibility and authority. Whatever lessons they draw about authority in general affect their comprehension of organizational power and possibilities for acting. Insofar as they unconsciously dwell on past conflicts with authority, they may turn to escapism and defensive ritualism, instead of taking initiative and exercising their ability. Eventually, the balance between satisfaction with the psychological contract and shame anxiety will determine whether bureaucrats can collaborate in solving problems or will scapegoat surrogates associated with them.

In order to promote their competence, workers must learn to recognize their assumptions about their organization and colleagues and be alert to conflicts between their thoughts and feelings and bureaucratic realities. There is no magical way to do this. Workers should try to ask themselves what broad aims lead them to take specific everyday actions. This questioning is particularly important, though difficult, when conditions seem unsatisfying or when actions have been unsuccessful. In addition, a telling cue to psychological issues is any sense that the organization, work, others, or one's own actions or reactions seem "strange." For example, events, issues, or situations may seem to be different than they overtly appear. Reactions may seem extreme or inappropriate. It may feel extraordinarily difficult to make a simple decision. One may repeatedly go over the same ground, never progressing. One may fixate on being angry or avoiding work. One may appear to do little about an evident problem. One may feel inept or stupid. These experiences are likely signs that unconscious assumptions are controlling organizational events. In all these cases, the purpose of self-questioning is for workers to figure out how much they themselves are responsible for situations and how much responsibility lies elsewhere.

In making sense of situations, workers should explicitly consider the following types of questions, probably already answered in negotiating psychological contracts:

> *Career and work*: Where do I want to be in the future? Where do I expect to be? What dreams do I have about work? What do I want to be able to do competently? What rewards do I expect from work? What do I want to be recognized for? How do I feel about work in comparison with family life? What makes me feel secure?
> *Job*: What do I want from this job? How is it related to career goals? What do I want to do competently in the job? Do I want to advance or just spend time? What do I

Table 4. Examples of Interventions

	Goals/Targets of Intervention	
Strategies/Domains of Intervention	Psychological Structure/Predicaments and Responses/ Problem-Solving Approaches	
	Tactics of Intervention	
	Diagnosis/Learning	Overt Action
Individual	Analyze own assumptions, expectations, intentions, interests, thoughts, and feelings about career, work, the organization, and relations with other; give particular attention to problem situations and "strange" situations.	Clarify expectations, and test assumptions; clarify ambiguities; modify expectations. Accept the irreducibility of some uncertainty and ambiguity. Accept the inevitability of mistakes, and attempt to learn from them. Become more visible to others; reveal interests and expectations. Change positions in the organization; change the organization; leave the organization. Consider both the substance and context of any problem. Distinguish between realistic and magical solutions for any problem.
Interpersonal	Clarify mutual assumptions, expectations, intentions, interests, thoughts, and feelings in relations with others; give particular attention to problem situations and "strange" situations.	Test assumptions about others' assumptions, intentions, interests, and actions by consulting with them; in particular, test assumptions about responsibility and authority for solving problems. Clarify expectations, and negotiate agreements on legitimate expectations.

have to give up in order to do the job well? Does the job give me control over things that are important?

Organization: What do I want from this organization? What do I have to give up to join the organization? How do I think of the organization—warmly, hostilely, or indifferently? How does it compare with my family? Does it permit me to work competently? Does it provide appropriate recognition and rewards? What satisfactions does my role offer me? In what ways does it confine me?

Other people: What relations do I want with others—for example, active support, cordiality, or indifference? What power do I want? What do I expect? What do I think and feel about supervision? Do I feel ashamed or guilty toward my supervisor? How do I

Table 4. Continued

	Goals/Targets of Intervention
Strategies/Domains of Intervention	Psychological Structure/Predicaments and Responses/ Problem-Solving Approaches

	Tactics of Intervention	
	Diagnosis/Learning	Overt Action
Organizational	Engage consultant to examine assumptions, expectations, intentions, interests, thoughts, and feelings affecting organizational mission; give particular attention to problem situations and "strange" situations	Experiment with cooperative activities; develop interpersonal and group competence through experiment and training. Support others' efforts, particularly those involving extraordinary initiative or risk, and minimize blame for any problems. Reduce organizational hierarchy; decentralize authority; delegate authority consistent with responsibility. Establish temporary problem-focused task groups. Individualize evaluation procedures; focus on workers' areas of competence, including their willingness to take risks with it. Encourage and reward reflective inquiry about organizational purposes, problems, and programs. Innovate technically to promote the organization's mission or to improve working conditions. Establish normal procedures for advising, generally applicable to consultants. Establish long-term relationships with consultants.

regard others—basically friendly and cooperative, or competitive and hostile? Do I feel I can trust others? Do I need others in order to work competently, or can I do without them? How do relations with others at work compare with relations with others in my family?

Levinson (1972) and Maccoby (1976) offer additional questions: there is no correct list.

Argyris and Schön (1974, 1978; Argyris, 1982) offer a formal method for organizing this inquiry by focusing on problematic encounters. They emphasize that difficulties arise not only when workers' interests conflict, but also when workers do not communicate their assumptions to others. In these cases, people imagine what others expect and cannot realistically assess the accuracy of their assumptions. In order to analyze problematic situations, Argyris and Schön suggest reconstructing transcripts, reporting what one said, how others responded, and so forth. For each overt statement, one should attempt to identify one's thoughts about speakers' tacit meanings or intentions, such as assumptions about relations between people in the situation, theories about how decisions should be made, evaluations of preceding ideas, reactions to criticism, responses to expressions of feelings, and covert efforts to get people to act in particular ways.

For everyone these tacit meanings fall into consistent patterns, what Argyris and Schön call a "theory-in-use," a set of assumptions that shape each person's relations with others. These assumptions, which are usually unstated and taken for granted, influence the ways in which people collect or ignore information about problems. For example, many people assume that others get hurt when challenged and, therefore, avoid asking difficult questions that might produce useful information. Many people also assume that others could not be trusted to collaborate and, therefore, parry others' questions and avoid sharing information. When several people in a situation hold these assumptions, they become mutually reinforcing and self-fulfilling, and people turn from solving problems to defending themselves. Argyris and Schön label this typical set of beliefs "Model I," summarized in Chapter 4. They show how workers, if they can recognize their tacit ideas about others, can begin to replace confining patterns of thinking and acting with more effective patterns.[3]

For example, the woman in Chapter 2 who is confused about organizational power might increase her effectiveness if she took some effort to look more closely at her own assumptions. Although she calls herself a "committed bureaucrat," she has difficulty understanding how people exercise power in her bureaucracy. Moreover, even though she knows she must figure this out to advance, she recognizes that she tends to block out thoughts about who exercises power. Thus she seems to act self-defeatingly. She might examine more carefully what career she envisions as a "bureaucrat." She might discover that her expectations of a bureaucratic role conflict with her career expectations. If she examined some of her meetings with her superiors she might find that she has mixed motives about asserting herself and refraining from any action threatening reprisal. By exploring these possibilities, she could start to assess the degree to which her dissatisfaction or apparently self-defeating action is a personal difficulty and how much of it is a plausible reaction to other people or bureaucratic structures. She then has to decide what to do next, but by analyzing her actions she is less likely to continue to act in unsatisfying ways.

Interpersonal

Interpersonal learning is possible when workers observe and discuss their interactions with each other. Each may learn about the other and about their relationship. In addition, each can test his or her assumptions about him- or herself by comparing them with the other's perceptions. Diagnosing interpersonal situations is important in discovering why some efforts may be unsuccessful or frustrating, some problems may not go away, or some situations may seem "strange."

Workers should organize their discussions around the two types of relationships in which they are involved—work with peers and hierarchical relations with subordinates or supervisors. Individual initiatives, past relationships, and organizational customs will affect who is likely to get together and what is open to discussion. Some people may be able to talk about certain topics more openly than others. Peers may be able to talk more easily with each other than a superior with a subordinate. Yet over time workers may learn how to learn together. If they can look explicitly at the following types of questions, they are likely to discover any major problems in their relationships and learn more about how to collaborate:

Purpose of the relationship: How much choice do participants have about being involved with one another? What benefits can they get from the relationship? What do they expect from each other? Do these expectations express personal feelings of caring or primarily impersonal role requirements? What do the people involved have to do to keep each other satisfied? What are the consequences of not satisfying certain others?

Interests: What are participants' interests? Are they compatible or competitive? Can conflicts be reconciled?

Who has authority over what in the relationship? In addition to formal authority, does anyone have informal authority related to intelligence, knowledge, personality, resource control, or something else? Do some people have more authority to enforce their expectations than others? How does the distribution of authority affect people's perceptions of or feelings about one another? Is the distribution of authority consistent with getting work done, or does it interfere?

Problem solving: What kind of relationship is necessary in order to solve organizational problems? Do the people involved have any interest in collaborating to solve problems? What have been their past experiences in working together? How might their respective strengths and weaknesses be put together in order to solve problems? Do current incentive systems provide any rewards for collective success, or do they emphasize individual performance? How could collective success be recognized and rewarded?

There is no single best way to examine these issues. Informal and occasional discussion may be easiest and effective. Workers can formalize their observations by jointly following Argyris and Schön's framework for analyzing problematic interactions. Woodcock and Francis (1979) also offer exercises for answering these questions. Significantly, however these issues are reviewed, any discussion that clarifies situations also changes them. For when bureaucrats know more about one another, their choices are less ambiguous, and they can make decisions on the basis of contextually more specific information. In addition, workers will not have as much need to add details of one another through unconscious transference. Further, administrators can bring formal authority to any such talks, with the result that they can convert new perceptions into formal decisions about authority and responsibility.

The senior federal worker in Chapter 3 who says "the biggest ambiguity was the anomalous relation of a deputy to his principal" offers an example of relationships based on preconceptions. He continually refrained from testing his assumptions about his boss' expectations or the balance of their authority. He concentrated on not challenging his boss and kept sufficient distance from him to avoid charges of threatening him, and now he believes that he was penalized for not taking available initiatives. Although he implies that his boss' preferences set firm limits, part of his uncertainty may have resulted from his own mixed feelings about asserting himself and following his ambition. If he had been willing to test his assumptions more explicitly over the years, he might have better understood what he and his boss expected of one another. By checking with his boss he could have assessed whether he was making prudent choices or cheating himself of greater opportunities. In turn, the boss might have learned that his deputy was willing to do more than he was currently doing.

Organizational

In order to understand how formal structures or pervasive norms affect problem-solving efforts, it is necessary to study an organization as a whole. Individuals, even small informal groups, cannot by themselves know all about these structures or norms or exercise significant control over them. Together, members of an organization have all pertinent information about how the organization affects them. If they compare their experiences, they can determine what is idiosyncratic and what is pervasive. Moreover, when members come together for this task, an observer can see collective patterns of behavior, some of which may be unconscious and beyond workers' everyday awareness. This makes it easier to identify connections between corporate structures and policies and workers' conscious and unconscious thoughts and actions. As with individual and interpersonal learning, the purpose of this diagnosis is to find possible causes for persistent substantive or procedural problems or to investigate "strange" conditions.

Members relate to organizations through groups. People work on tasks in groups, which may be formal departments or divisions of one or more departments. Formal authority relations constitute groups of people accountable to one another, who meet occasionally for assignments or evaluations and who think of each other. In addition, people tend to conceptualize authority in terms of past relationships and unconsciously create "groups" combining present and past actors. These "groups" may overlap work groups; they are probably closely related to networks of authority relations. Organizational diagnosis should focus on the activities, meanings, and connections of these groups.

Because of the complexity of organizations, their internal conflicts of interest, and many opportunities for misunderstanding and distortion, an outside consultant is crucial for learning. Chapter 7 identified workers' common misgivings about and resistance to consultation. Nevertheless, an outside advisor may offer a disinterested perspective that outweighs these potential problems. Further, a consultant may be able to help members see that they resist learning and why. In order to work legitimately, consultants must be hired by top management. However, in order to facilitate organizational learning, consultants must convince subordinates that any investigation will be disinterested, open, and appropriately protective.

Both administrators and subordinates may be ambivalent about examining their organizations. At the same time they may want to know more about sources of problems and fear what they may find out. Administrators may worry that analysis could lead to criticism and blame, reveal problems they could not easily solve, and end in dissent and revolt. Subordinates may believe that open criticism will lead to punishment. Both may prefer to live with moderate strains than risk more serious crisis. Consequently, crisis is often the main impetus for organizational study. But even then managers may attempt to analyze their organizations while holding back penetrating questions, defining apparent problems in technical terms, as difficulties in production, rather than in human relations. Particularly in service organizations, separating the two makes little sense.

In order to avoid both these risks—postponing learning until a crisis or defining problems in technical terms—administrators should institutionalize consultation procedures that examine human relations issues in conjunction with technical issues. For example, most organizations have regular planning cycles concerned with programmatic challenges and opportunities. Because production depends on collaborative relationships, managers should formally include human relations in routine planning analysis. Analogous to distinctions between five-year plans and annual plans, organizations may schedule major studies every few years while implementing smaller efforts annually. The institutionalization of organizational diagnosis may not only prevent crises, but its separation from specific crises or problems sets it off from individual interests and promotes more openness.

Any study should focus on significant organizational groups, such as groups of administrators, departments, work groups, and subordinates with their hierarchical superiors. Many of the questions discussed previously are useful here as well. Whom consultants interview and what they ask depends on the organization and what its members are willing to talk about. After interviewing workers, consultants should bring people together to discuss their responses collectively and to examine their reactions to what they have said. Consultants should also observe people at work, to compare what they say with how they act.

There is no ideal method for organizational self-study. Different approaches address different issues. In the end, an organization needs to select a consultant who works well with members on matters they want to explore. Levinson (1972) has prepared an insightful, detailed guide to organizational diagnosis. His book, addressed to consultants, can help administrators as well, in describing purposes of diagnosis, ways of conducting it, data to examine, questions to ask, methods of analyzing data, and means of discussing findings. He provides examples demonstrating the ties between technical and psychological (or human relations) issues.

Kets de Vries and Miller (1984) offer a more explicitly psychoanalytic framework for organizational diagnosis. They inventory unconscious phenomena that create problems, ranging from individual to dyadic, small group, departmental, and whole organizational conflicts. They offer categories for analyzing issues in psychological contracts, transference and other relations between superiors and subordinates, small group and departmental cultures, and overall organizational styles. They illustrate their intervention method with individual, interpersonal, and organizational problems.

Argyris and Schön (1978; Argyris, 1982), focusing on cognitive processes and communication, have adapted their methods to use with organizational groups. They describe how consultants may help workers identify tacit assumptions about human relations in dominant organizational norms. Case studies show how their framework elucidates common defensive assumptions, in which members resist discovering differences of opinion and prefer to act on faulty information. Argyris and Schön describe some success in helping bureaucratic workers to learn more effective ways of thinking about problems.

Mitroff (1983; Mason & Mitroff, 1981) has developed a method for identifying "stakeholders," which affect thinking about organizational policy, procedures, definitions of problems, and solutions. Most often, people think of the influence of social, or political, actors, for example, clients or customers, staff, and funding agencies. Mitroff introduces various psychological "stakeholders," including bureaucrats' unconscious wishes. He shows how workers with different personality types prefer and attempt to bring about different types of organizations. Thus programmatic decisions reflect not only technical judgments, but also workers' psychological stakes in specific types of organizations. Mitroff presents a method for helping members identify connections between their conscious and unconscious interests and organizational policy.

Chapter 3 mentioned a federal policy analyst who reports that many of his peers lack assignments and are not looking for any. If he is right, organizational diagnosis would benefit his agency. A consultant could help determine whether what the analyst describes is widespread, and if so, how different members think of it. By interviewing people confidentially, a consultant could begin to find reasons for the behavior, perhaps ranging from deliberate malingering to unconscious reactions to autonomous authority. The study might show that many well-paid workers feel angry that they have little authority or that no one recognizes them and that many choose escapism as a remedy. If members were willing to discuss these findings, they could provide additional information with which to analyze organizational incentives, the roles of individual administrators, perhaps even their own complicity in avoiding authority. This critical examination could be the beginning of efforts to solve some of these problems.

OVERT ACTION

The following examples illustrate actions that could improve organizational psychological structures, mitigate predicaments, make responses to predicaments more reasonable, or directly improve problem solving. The examples are organized by the domain in which they intervene.

Individual

Clarify or Modify Expectations; Change or Leave the Organization

Workers may fail to work out satisfying psychological contracts because their expectations are nebulous or unrealistic for a specific organization. Sometimes workers have vague expectations because they have given little thought to what they want or might get. On the other hand, some bureaucrats have ambiguous expectations because they have conflicting goals and cannot choose among them. For example, the policy analyst in Chapter 2 talks vaguely about wanting to exercise power. Her vagueness is the result not of insufficient thinking, but of a great deal of thinking, much of it unconscious, about how both to exercise power and to protect herself from punishment for being aggressive. She may believe that if she thinks about power only vaguely, she can protect herself from retribution.

In simple cases, workers can clarify ambiguous expectations by studying their organizations more, thinking more about their preferences, and making choices. In this case, for example, the policy analyst could attempt to understand her difficulties in thinking definitely about what she wants. She might recognize her conflicts about asserting herself. She might conclude that her fears are unrealistic and begin to develop more specific plans for exercising power in the organization. On the other hand, if she felt that she could not drop her fears of punishment when power is exerted, she could modify her expectations. Instead of thinking about influencing decisions overtly, she might decide to concentrate on performing cogent policy analyses and through them influence policy indirectly.

Alternatively, some bureaucrats may conclude that, although the organization does not satisfy their expectations, their needs are important and reasonable, and they may think about changing the organization to get more recognition, act more competently, exercise more power, or the like. Whether this is reasonable depends on a worker's position and power in the organization.

Finally, bureaucrats may conclude that they cannot change their organization and can find satisfaction elsewhere. This assessment can be reasonable, although some people repeatedly quit jobs, each time with a plausible excuse, in order to avoid anxiety about getting too close to colleagues, succeeding, or something else which makes extended affiliation uncomfortable. Thus this policy analyst might decide that she could not work competently in her organization. In fact, at the time of the interview she was studying law, and she left the agency about two years later.[4]

Become More Visible to Others; Reveal Interests and Expectations

Workers' reactions to one another are based on a combination of informed perception and speculation, which is often unconscious and frequently includes transference of assumptions from earlier experiences. If these assumptions are unrealistic, workers may expect fantastic rewards from the organization, build up immobilizing predicaments, respond ineffectively to them, or avoid the instrumental requirements for solving problems. Those who have little direct information about significant others have little choice but to fill in details from imagination, particularly when formal structures limit contacts between supervisors and subordinates. However, superordinates often have discretion to reveal more of their interests than they normally do. If they did so, they could give subordinates more realistic information about their expectations, and they could reduce subordinates' unconscious needs to transfer inappropriate assumptions from elsewhere.

Jones, for example, complains that her boss is rarely visible and often vague. If she is right, by making his own position clear, he could help subordinates curtail unconscious fantasies about him and the organization and invest more in formal tasks. However, as other administrators, he might resist revealing his expectations if he used ambiguity to worry subordinates into making extra effort. He might also cherish ambiguity as a way of avoiding responsibility for controversial actions. Thus in making himself more visible, he would have to give up these benefits, but he could gain subordinates' commitment to organizational goals.

Consider Both Substance and Context of Problems

Whenever bureaucrats get to work on problems, they (or their advisors) often run into trouble when they focus on the apparent substance of a problem while ignoring its context. For example, someone may think a great deal about income maintenance issues raised by a new proposal without giving much attention to politicians' opposing positions on the issues, changing public opinion on work and welfare, an agency director's needs for a position on the issue that protects budgets and staff, the director's fears of taking a stance similar to one on which he was defeated in the past, or an intensely personalized conflict over options within the agency. Bureaucrats and their advisors do not necessarily have to solve these problems, but unless they recognize their impact on substantive issues, they may define a problem in such a way that its "solution" does not fit politically or organizationally.

The case of Simon provides an example of the importance of scrutinizing both aspects of issues. If his client, the director, had been willing to consider how his political agenda both contributed to council problems and blocked solutions for them, he might have given Simon more explicit instructions. Instead, assigning Simon to work with his deputy reenacted some of the problems, by ambiguously assigning responsibility without authority. He might have recognized the conflicts with his deputy and asked Simon to meet periodically with him, to avoid split loyalties in analyzing the program. Because the director was an astute politician, he possibly understood but chose to ignore these issues, in the belief that disclosing them to Simon would jeopardize the director's control.

Whatever the director thought or did, Simon should have thought more about relations between the director and his assistant. Then he might have collected more information about the director's political relationships. In addition, Simon might have taken the initiative to maintain closer contact with the director, to apprise him of findings, and avoid even accidental alliance with the deputy against the director. He might have been able to focus more clearly on the director and deputy as a unit, albeit one with conflicts, rather than ignoring the conflicts and inadvertently playing into the assistant's unconscious efforts at scapegoating.

Distinguish Between Realistic and Magical Solutions

At any time, to forestall scapegoating, bureaucrats and their advisors need to identify what a realistic solution for a problem would accomplish. They need to recognize the possibility that instrumental problem solving gets diverted into magical thinking and must be aware of the signs of magical "solutions." In the case Simon worked on, the director might have concluded that a realistic solution to the council problems required not simply new assignments to councils, but also some redistribution of authority. He might have acknowledged that his wishes to control the councils, while satisfying, permitted only magical solutions to council problems. Even if he had not come to this conclusion, Simon could have given more attention to the symptoms of magical thinking. He might have recognized the assistant's emphasis on her split with the director as a sign of interest in scapegoating. By raising these possibilities, he could have gotten the director and deputy to think more realistically about alternatives.

Interpersonal

Develop Interpersonal Competence

As workers attempt to discuss their needs with others, come to agreement about legitimate mutual expectations, or cooperate in carrying out assignments, they may sense that hidden agendas or covert assumptions keep them from understanding one another and collaborating effectively. Often bureaucrats complain that they have difficulty talking to someone or persuading others to take their ideas seriously. Some of this puzzlement arises from workers' ambivalence about acting assertively, but some stems from their lack of interpersonal skills in assessing the thoughts or feelings of others, analyzing underlying themes in group relations, couching ideas in ways sensitive to others' interests, and responding to tacit, as well as overt, concerns.

Bureaucrats would be more productive if they could recognize unconscious events typical in work groups, learn to analyze them, and plan instrumentally despite any mixed feelings about belonging to a group. Interpersonal competence for a work group requires that all individual members be able to work with the others. Those who want to get past hidden issues may make deliberate, self-critical efforts to think about, discuss, and resolve some of these issues. Some individuals enter organizations more competent in these respects than others; they may teach others both explicitly and by example. In addition, work groups may collectively seek training in communication and group problem solving. Training is most likely to be effective when management recognizes its value, pays for it, and supports its use. Alternatively, members of a work group may choose to get training for themselves, or individuals may do so.

Many training methods may be helpful. Some give more overt attention to unconscious aspects of groups; some present more explicit models of leadership and rational action. For example, group relations conferences, developed from the work of Bion and the Tavistock Clinic, are explicitly psychoanalytic and are most useful for making participants aware of the ways in which people draw unconscious assumptions about one another in groups (see Astrachan, 1975; Colman & Bexton, 1975; Colman & Geller, 1985; Gustafson & Cooper, 1979; Klein & Astrachan, 1971; Redlich & Astrachan, 1969; Rice, 1965; Rioch, 1970). These study groups focus on how members' statements and actions express unconscious beliefs about relations with each other. Consultants making observations on covert group events help members articulate their tacit thoughts and feelings about the roles of members, their reasons for coming together, and how work could be accomplished. Balint (1954, 1957) developed a variation of this model to familiarize physicians with unconscious dynamics in relations with patients. In these groups consultants are more active in teaching members about how to interpret and work through typical unrealistic assumptions (Gustafson & Cooper, 1985).

Training groups, or T groups, created by Lewin and the National Training Laboratories, examine recurrent group phenomena nonpsychoanalytically. Groups focus on teaching collaboration by identifying how participants create obstacles to cooperation and showing how to overcome the obstacles (see Benne, Bradford, Gibb, & Lippitt, 1975; Bennis, Benne, Chin, & Corey, 1976; Bradford, Gibb, & Benne, 1964; Schein & Bennis, 1965). T group leaders take more active roles than in the other groups and are the most involved in modeling and teaching leadership and problem-solving skills in ongoing situations.[5]

Any of these methods offers people working together the opportunity to learn new ways of thinking about and acting toward one another. How much they learn and change depends on their own goals and fears, whether they want to collaborate, and whether they are anxious about taking overt responsibility for decisions. Learning and change depend also on how much innovation organizational structures and administrators will permit or accept, for example, whether administrators are comfortable with more active work groups. In both respects, what seems difficult initially may become possible if workers are willing to risk making changes that could increase their competence and satisfaction.

For example, the federal analyst in Chapter 3 who emphasizes that he always looks out for himself complains that he has difficulty getting cooperation from colleagues when he assembles project teams. If he could get group training for his team members, they could learn to recognize and discuss some of their concerns. For example, some may avoid any responsibility because they fear being shamed by supervisors. Others may feel embarrassed and resentful working for team leaders in lower job grades. Still others may feel slighted about having their past ideas ignored. Although the analyst cannot restructure organizational incentives or change anyone's personality, he can work with team members in making arrangements that avoid some of these finally acknowledged obstacles to cooperation.

Negotiate Agreements on Mutual Expectations

Whenever people work together to solve problems, whether in ad hoc project teams, permanent task groups, director-deputy dyads, or formal consulting relationships, they need to agree about what they can expect from one another.

This agreement is central to psychological contracts. If a worker can get a supervisor and significant peers to recognize his or her expectations as legitimate, satisfaction is more likely. Even just talking directly about mutual interests in attempting to reach agreement is useful. In many cases, coworkers may conclude that each other's wishes, though never before mentioned, seem reasonable, and they may agree to try to satisfy them. In other cases, some may recognize others' needs as legitimate but indicate that they cannot meet them. Even though this conclusion may be disappointing, overt discussion and recognition of the needs may improve communication and working relations. Coworkers may find times when their interests, even if they seem reasonable, conflict, perhaps just as they had imagined. However, this discovery too may be beneficial. When workers identify conflicts specifically, they may be able to negotiate some to conclusion. Finally, there may be instances in which workers find that no one regards their needs as legitimate. Although this discovery is unpleasant, recognizing this may lead some workers to reassess their position and eventually find ways of working more effectively.

More generally, efforts to discuss and negotiate interests unconsciously affect an organization's psychological structure. The more that workers talk about their beliefs and expectations, the more they enable each other to form accurate pictures of one another. For example, even if a subordinate cannot get a supervisor to agree to certain wishes, the negotiation may help each understand the other better and react more realistically. Both are less likely to need fantasy material to fill in an image of the other. In addition to reducing the realm of the unknown, negotiations may diminish inequalities in status or power. As a result, even when bureaucrats do not know everything about those on whom they depend, they will be less likely to think anxiously of them as omnipotent parentlike figures.[6]

These negotiations are especially important when people working together do not know anything about each other's past or work habits. For example, consultants and clients need to agree about their respective responsibility and authority. Explicit discussion minimizes the chances that a client who would like to be generous with authority or to shift blame for failure gives the consultant mixed messages about authority. Similarly, an advisor is less likely to assume authority he lacks simply because he has great confidence in his ideas. By talking about their expectations and wishes, clients and consultants can negotiate agreements, reducing the probability of either scapegoating or defensive ritualism.

Interpersonal competence will help bureaucrats identify their tacit interests in relation to one another and develop agreements about mutual expectations. Fisher and Ury (1983) argue that negotiations generally are more likely to be successful when parties accept each other's interests as legitimate. They present negotiating techniques based on four strategies: (1) parties should distinguish substantive problems from human relations problems and treat them separately; (2) they should focus on interests, rather than specific transitory positions; (3) they should generate many possibilities before making a decision; and (4) they should base the result on an objective standard.

The federal civil servant who worried about climbing over his boss serves as an example of how explicit negotiations could have improved working life. As a deputy, he might have talked with his boss about his career goals, including thoughts of developing new competence, acquiring influence, and gaining status. The boss might have indicated how realistic he thought these expectations were, which ones he could support, and which might be revised, and in turn he could have said more about what he wanted from a deputy. For instance, the boss might have wanted public loyalty on some issues while permitting disagreement on others. He might also have recognized that his deputy should move on after a few years, and he might have offered to support the deputy in advancing if he served loyally in the meantime. Even if the deputy were not pleased with all his boss' stated expectations, such a frank discussion would have eliminated a lot of the ambiguity that made him anxious whenever he considered taking initiative.

The consulting case with Simon is an example of a consultant who tries to be agreeable but avoids discussing disagreements. He could have said more to his clients about his private concerns. He should have pointed out to the director difficulties in working only with the assistant. He could have talked explicitly with the assistant about risks of his getting caught between her and her boss. He should have asked more about why she wanted to exclude the director from the chairpersons' meeting. Similarly, he should have directly questioned why she was offering him the previous consultant's letter. In each instance, he might have helped the director or the assistant clarify their expectations of him. He might also have helped them understand how conflicts between them could jeopardize the consultation. In any case, he might have gotten them to take more responsibility for the consultation.

Organizational

Reduce Hierarchy

Perhaps the most widely advocated structural reform of bureaucracy is flattening the organizational hierarchy (see Denhardt, 1981; Fischer & Sirianni, 1984; Harmon & Mayer, 1986; Hummel, 1982; Thayer, 1981; White, 1969). Proponents emphasize gains in democracy. As more people participate in deliberations, they can each contribute relevant information, and decisions will be more knowledgeable. In addition, participants will support decisions and take responsibility for implementing them. A number of analysts argue that these conditions are particularly important for service work, including planning and administration, because, unlike manufacturing, it involves many unique cases and requires discretionary judgment rather than routine procedures (see Galbraith, 1977; Katz & Kahn, 1978; Mintzberg, 1979; Perrow, 1970).

These results depend on whether new formal structures evoke new unconscious responses. Flattening the hierarchy can temper the psychological structure by permitting more satisfying, less threatening relationships. Insofar as workers are formally more equal, they are less likely to regard others as oppressors and to react hostilely, aggressively, anxiously, and defensively. Both real threats and unconscious embellishments of power differences may diminish. When social distance is less, superiors and subordinates may more freely discuss expectations and substantive problems. Thus members may clarify some ambiguities and reduce the situations in which they unconsciously draw threatening images of one another.

Flattening the hierarchy can also reduce the distortions produced by autonomous authority. Diminished distance between those who do exercise authority and their subordinates will make authority more visible and accessible, thus less autonomous. Regardless of the personality or style of people with authority, subordinates will know more about them and be less likely to make vastly inappropriate assumptions about them. At the same time, a shallower organization reduces subordinates' dependency. Even if they find their remaining dependency shameful, their greater control should make them less anxious than previously. Consequently, when subordinates do make unconscious assumptions about administrators, they are less likely to begin with intense shame and efforts to protect themselves against it. They will have a better chance of realistically assessing problems and others' ability to help solve them.

The situation with Jones is an example of a bureaucrat who knows about significant organizational problems (such as blocks to staff participation) but whose position is so far subordinate to formally designated professional staff that her superiors rarely ask her what she knows. As a result, she conjures up a play world in which her knowledge counts. Smith has more formal authority but still chafes at limits on what he can do, and he leaves to create an organization in which he makes the rules. The policy analyst who emphasizes looking out for himself complains that he cannot get analysts with higher job grades to work for him on project groups. He believes that hierarchical distinctions inevitably interfere with sharing knowledge. If any of these organizations were less hierarchical, staff might be both more satisfied and more likely to collaborate.

Nevertheless, formal changes that apparently benefit many may entail unconscious costs and arouse resistance. Analyses of failed bureaucratic reform often point to administrators' resistance to losing their accustomed control. More significantly, other accounts (for example, Crozier, 1964; Gouldner, 1964; Menzies, 1975) call attention to subordinates' opposition to changes that would apparently give them more authority. Despite complaints about lack of authority, some find their hierarchical lack of power reassuring. Rarely encountering those with authority, many subordinates feel secure, for they do not have to worry about how others use their authority, how they might protect themselves, whether they could negotiate without being seduced or crushed, or how they could take responsibility for others whom they might defeat. Thus many bureaucrats may at least initially feel anxious about these structural reforms giving them authority and forcing them to confront others with authority. Perhaps they would react to new, formally imposed closeness by creating new, informal means of distancing themselves from making decisions. Individuals must recognize their unconscious pleasure in subordinacy and dependency before they can freely choose to accept more formal authority. Clearly, change in all domains is linked.[7]

Establish Task Groups

Organizational reformers consistently advocate temporary task groups (see Biller, 1973; Michael, 1973; Miles, 1964; Schön, 1971). When a problem arises, a manager may convene individuals specifically skilled in working on that problem, support this group so long as the problem persists, and disband the group when they have solved the problem. This arrangement improves an organization's competence by substituting specially constituted groups for standing departments or divisions. At the same time, task groups offer practitioners the opportunity to work competently.

In addition, task groups offer unconscious benefits in modifying the organizational psychological structure. The opportunity to work competently satisfies one of a member's primary expectations. The possibility of working closely with others similarly interested in solving a problem both reduces formal differences in authority and requires intense consultation. Morever, members must share responsibility for solving problems. At least temporarily, task groups thus eliminate primary sources of organizational anxiety.

The policy analyst who reports that only some staff members have projects to work on points to the mixed attractions of task groups. He says that he enjoys the opportunity to work competently in the small group he coordinates; some of his colleagues envy him. At the same time, he says, others are relieved not to be involved in project groups because they are free from responsibility. Although, as he implies, some may be avoiding realistic risks of blame, others shun responsibility for the same unconscious reasons that workers may resist flattening the hierarchy. In order for temporary task groups to be effective, bureaucratic workers must confront conflicts about taking initiative and be willing to risk the commitments necessary for collaboration.

Individualize Evaluation Procedures and Focus on Competence

Along the same lines, proponents of pluralistic evaluation argue that workers should be measured on their willingness to do what they can do well, rather than simply their performance of routine responsibilities. Such a formal arrangement would simply extend interpersonal negotiations over psychological contracts. Schein (1978) describes ways in which supervisors may develop relevant evaluation standards.

Formal recognition of workers' abilities makes psychological contracts more satisfying and reduces the shame associated with autonomous authority. Consultation about evaluation criteria gives subordinates more control over decisions about organizational rewards. It requires supervisors to state their expectations and enables subordinates to react more realistically. In general, subordinates are less likely to avoid talking with supervisors out of fear of being shamed for something they could not do or for doing something wrong. Consequently, subordinates will be less likely to feel angry at supervisors or feel guilty about being angry.

Nevertheless, workers who unconsciously have stakes in their subordinacy may oppose change. For example, many people find any evaluation shameful and would prefer to keep it as impersonal as possible, even if this means sacrificing the possibility of individualized assessment. Rather than risk negotiating closely with powerful superiors and taking responsibility for setting standards, some would prefer to be judged, even inappropriately, by distant superiors. Primarily concerned about avoiding shame, they may cherish the possibility of alleging that their evaluator could not have understood their work, because they fear that any fair evaluation would shame them.

Jones illustrates this common ambivalence about evaluation. She ridicules formal job descriptions that have little to do with actual responsibilities. Because she does not trust her supervisor to evaluate her fairly, she resists impulses to correct the formal description. Instead, she says, she will leave matters so that, if her supervisor ever tried to evaluate her, he would have to confess that he had no way of knowing what standards to use. However, as she acknowledges, so long as there are no definite standards, she cannot get realistic praise either. If she were willing to risk approaching her superiors, she might negotiate what would amount to individualized evaluation criteria that could give her positive recognition and satisfaction. Her reluctance to take this initiative illustrates many workers' hesitations about pushing for or supporting individualized evaluation.

Innovate Technically

Organizational effectiveness entails not simply satisfying human relations, but also appropriate materials and rational procedures. The two realms are connected. Instrumental tasks focus workers' energy; when prescribed procedures do not engage or solve problems, workers turn to thinking about one another. Their fantasies may take relatively free rein, conjuring up imaginary threats from others and designing defenses against them. Thus without an instrumental focus, staff members not only are unproductive, but also may create a frightening unconscious world that pushes them further away from the formal organization.

Hence when organizations lack effective technologies for their missions, or when missions do not fit the environment, both technical and human considerations call for re-evaluating the mission and innovating technically. Innovation may involve designing new equipment. Often, particularly in service organizations, innovation entails directing staff to collect new information about old or recent problems, consult with other people, use different models for analyzing information, and disseminate information to new clients or audiences. The overt result is to improve organizational effectiveness. This added potential for success may unconsciously help workers focus their thinking on instrumental actions, instead of dramatic play, ritualism, or escapism.

For example, Jomes refers to technical failure when she talks about administrators who do not understand managing, job descriptions that do not fit workers, and purely ritualistic regulations. Her agency recently underwent a politicization in what traditionally was a highly professional organization, and political changes led to a rapid succession of several directors, each with short-term political interests and little familiarity with the agency's history. Politicization of programs introduced new missions while making old programs obsolescent. Many old procedures, as Jones complains, no longer lead anywhere. She turns to dramatic roles for several reasons, but this technical breakdown is a major impetus. Under the agency's current administration, technical innovation may be less important than political management, but it could give more focus to workers' energies, which, as in her case, go into activities less and less tied to organizational purposes or individual competence. The longer the organization continues on it current course, the more members such as Jones develop investments in their alternative activities, and the more they will resist technically reasonable reorganization.

Establish Normal Procedures for Advising

Technical obsolescence is a recurrent organizational problem. So are mismatches between tasks and staff, misunderstandings about responsibility and authority, and confusion of realistic and magical actions. However, bureaucrats like to assume that any problems are extraordinary, unanticipatable events. Because they cannot be prevented, no one can be responsible for them. Consistently, bureaucrats act as if consultation is an unusual event. Thus organizations rarely have normal procedures for advising, and each client and consultant must recreate conditions for collaboration. They are left to struggle with personal and organizational ambivalences about advising as if these, too, were unprecedented.

Instead, administrators should recognize that typical uncertainties or ambiguities periodically create needs for consultation. They can designate people within their organization as experts available for consultation in specific fields. In addition, formal policy can state conditions under which certain workers have the authority to request consultation. Some might have authority to request assistance from others in the organization, including the designated experts; others would have authority to engage outside consultants.

Formal recognition of recurrent needs for advice would help reframe organizational views of problems. If they are treated as normal, rather then extraordinary, events, workers may think less of blame and more of learning. In addition, regularization of advice seeking may reassure workers that they will be able to get assistance when they have problems and reduce their anxiety about being shamed. They will be less likely to conceal difficulties until they become crises. Together, these changes can reduce unconscious pressures toward scapegoating consultants.

The county executive's human services director, for example, presented the council's lack of productivity as an isolated problem. Council members apparently created it, and it had little to do with his overall management style. Just as he regarded this problem as annoying and wanted it gone quickly, he seems to have felt similarly about Simon. Significantly, the director concealed a past consultation from Simon and tacitly instructed the deputy not to discuss it. As a result, not only was Simon unable to learn some useful lessons about working with the director, but the deputy had no guidance in consulting. Consequently, she treated Simon as an unique advisor, and she created fresh conditions for using him in the context of her current conflicts with the director and mixed feelings about him. If the director had openly acknowledged his occasional needs for consultants, he could have reviewed his preferred ground rules with the deputy. He would thus have set her assignment in an organizational perspective and given her guidance and support in carrying it out.

These examples illustrate the range of interventions that can help bureaucratic workers interpret and change problematic conditions. Although some interventions require top administrative decisions, individuals or small groups can take other steps themselves. What these interventions have in common is the recognition that organizational psychological dynamics affect members' abilities to solve problems. The

interventions assume that workers are more likely to be effective when they acknowledge their tacit thoughts and feelings about work and deliberately choose their goals,
their relations with other people, and their work procedures.

CONCLUSION: TOWARD RESPONSIBILITY

Examining organizational psychological dynamics offers bureaucratic workers the
possibility of responsibility for their actions. Analysis makes explicit many interests
that normally remain hidden. It illuminates the complicated, shifting conflicts
among these interests. Perhaps most important, it helps explain how stronger
interests are able to maintain themselves against weaker, how the weak attempt to
defend themselves, and how each may collude in arrangements from which both
suffer.

Psychoanalytic investigation offers bureaucrats the possibility of reclaiming
action. It points out the ways in which workers contribute to predicaments they
attribute simply to the dominance or treachery of others. Although power in bureaucratic hierarchies really is unequal, workers may find that they have exaggerated or
increased the disparities as ways of avoiding the uncertainties of collaboration. By
discovering their own complicity in their powerlessness, workers have the chance to
make new choices, attempt to act powerfully with others, and take responsibility for
these explicit decisions and their outcomes. By reconsidering ways in which they
have attempted to defend themselves against the risks of exercising power, workers
may enact more satisfying psychological contracts and help organizations solve
problems.

Many bureaucratic arrangements serve as a defense against anxiety. Some
anxiety is inevitable in human life, and some defenses are helpful in permitting a
relatively uncomplicated conduct of everyday affairs. If bureaucracy did not block
certain impulses, then people would create alternatives to do so. However, this book
has focused on ways in which bureaucratic organizations, offering employment,
perhaps tacitly promising security from anxiety, nevertheless have given rise to new
anxiety and led to the erection of new, ever more complicated defenses. It is not clear
that this anxiety and these defenses must be part of work. The book is an invitation
to an emancipatory experiment, in which bureaucrats may attempt to discover what
they have unconsciously chosen and may consciously choose what to retain and
what to replace.

Psychoanalytic study of organizations offers workers political freedom. Interests and conflicts may be made explicit, and power may become visible. This new
insight can help distinguish conflicts of interest from clashes that arise from misunderstanding. Then bureaucrats, more clearly recognizing their interests and better
able to choose and design powerful actions, may better control what they do and
how they do it. Insofar as workers learn to collaborate, they may change their
organizations to permit effective work.

However, the new possibility of organizational politics points to likely resistance to organizational study. Those who currently benefit from covert organizational relations are likely to resist visibility. Many of these are the powerful. But even
those who regard themselves as weak may resist these changes. Sociologists frequently excoriate or praise bureaucracies as instruments of social control; this book

has examined the ways in which they are instruments of psychological control. In fact, many of the controlled are gratefully so, preferring bureaucratic domination to the chaotic worlds they imagine otherwise.

The preceding discussion of interventions suggests that workers may have fairly good reasons for resisting not only re-examination of the accommodations they make, but also any consideration of alternatives. Even though few of these interventions depend on psychoanalytic insight or justification, all affect people's unconscious thinking and feeling about their work and organizational relationships. Although people may gain refreshing competence, they stand to jeopardize a tangle of makeshift security arrangements.

It is uncertain how efforts to incorporate psychoanalytic insights into bureaucracy would proceed. In individual psychoanalysis, people resist psychoanalytic insights, even though they offer freedom and responsibility. There people incorporate the insights into ongoing conflicts, using them in support of some interests against others. Whether organizational actors would respond analogously is unclear, because social conflict is more than simply intrapsychic conflict writ larger.

The language of psychological dynamics is dramatic. It is not the customary rational language most planners and administrators favor. Rather, it is the grammar, if not exactly the vocabulary, of conventional politics, which many bureaucrats oppose. Perhaps they will not respond to the language, but perhaps they will note that the aim of its discourse, nevertheless, is learning, just like rational problem solving. Learning and resistance will set the markers for a complicated scenario of more psychodynamically sensitive bureaucratic work and thinking.

Before this scenario can be written or enacted, three types of research questions about the relations between individuals and organizations must be answered. The first questions are conceptual and empirical. They entail defining more precisely the ways in which individual dynamics, including individual problems, affect organizational dynamics and problems. This book has described some typical, predominantly problematic responses to the psychological structure of bureaucratic organizations. More needs to be known about variations, both in psychological structures and in individual responses, particularly where some of those variations are associated with greater effectiveness. It is important to discover organizational psychological structures more benign than those found here.

With regard to individual variations, it is essential to try to distinguish and characterize groups of people who appear to act with different measures of effectiveness. Some people may generally be more successful than others, perhaps because they have fewer conflicts about being competent, getting recognition, and exercising power. On the other hand, as LaBier (1984) argues, effectiveness may depend on a good fit between individual personalities, with all their conflicts, and organizational situations and role expectations.[8] It is important to explore the variations that distinguish competent from ineffectual responses to organizational problems. Although it is clear that bureaucratic structures serve individual defensive needs, more must be understood about the ways in which individuals may tacitly choose, design, maintain, or modify structures. In particular, it is important to know more about how different combinations of individual personalities combine to create different organizational psychological structures.

These questions raise methodological and practical questions. Simply, it is uncertain to what degree it is possible to study individuals' unconscious thinking as part of organizational study. Outside of organizations, intensive study of individuals may reveal a great deal about them, but relatively few workers choose psychoanalysis, and interactions in organizations involve more than can be learned from the sum of individual investigations. Within organizations, workers may regard overt examination of their unconscious thoughts as esoteric and extraneous to organizational analyses of productivity and effectiveness. In addition, they may resist this examination as intrusive, messy, threatening, and potentially uncontrollable. Subordinates, particularly those who feel estranged from or anxious about their bosses, are unlikely to agree to any investigation that risks exposing them to scrutiny and possible reprisals from superiors.

Broadly, many bureaucrats fear that, whether or not organizational study uses psychoanalytic terms, it will unearth problems that investigators then leave unresolved. Particularly because psychoanalytically informed methods for studying organizations are new, anyone using them must work closely with members in explicitly setting the limits of analysis and delineating study methods. Thus workers may share responsibility for the study and its consequences. As a result, they may help establish procedures for bringing to light what they are willing to reveal.

Finally, it should be clear, these methodological questions raise ethical questions. It is important to identify the risks to bureaucratic workers when the diagnosis of organizational problems includes the study of individual personality. Any research should adhere to an explicit contract involving not only top management but anyone who is interviewed or observed. In addition, the study, though it may be stressful, should not end in harm. This is particularly important because negative consequences are not simply distress or anxiety, but loss of employment and income.

More subtle possibilities of harm require attention. There are risks that the study of workers' motives will become incorporated into prevailing power relationships, that managers may exploit new information about subordinates in order to control them. This risk is particularly serious because manipulation of workers' unconscious lives is less likely to be visible or contestable than more conventional means of organizational control. There is also the risk that disagreement, normally a subject of bureaucratic politics and argument, will be reinterpreted as opponents' mental instability. Differences, generally the subject of confrontation, could become the basis for recommending or requiring psychotherapy. A central ethical question concerns the degree to which managers, who have responsibility for organizations, should have responsibility for individuals as well.

The psychoanalytic study of bureaucratic organizations is a bold experiment. It entails risks and uncertainties. Yet there is an undeniable argument for making the effort. Planners, administrators, analysts, and many others may discover the possibilities of work in organizations. They may find ways to take responsibility, act competently, and give and receive recognition. These are the conditions of a psychological contract in which both workers and organizations profit.

NOTES

1. In general, planners tend to avoid examining their motives. When they describe their work, they are much more likely to consider the external aims of their actions than to examine themselves as thinking, feeling actors (Baum, 1983a). This orientation resembles the defense

mechanism of "isolation"—wherein people detach conscious thoughts from associated unconscious feelings—by separating the practitioner's conscious calculations from any unconscious intrusions (Freud 1936; rep. 1963).

2. Lasch (1978, 1984) offers two complex hypotheses about how modernization, including the technological fragmentation of work and commoditization of everyday life, has affected bureaucracy. First, it makes bureaucratic administration seem both more reasonable and more efficient than overt politics as a means of societal governance. Second, people react to these societal changes by developing narcissistic personalities, which are ideally suited to the bureaucratic manipulation of interpersonal relations. For a general assessment of the influence of societal structures and norms on institutions and individuals, see Fenichel (1945, chaps. 20 and 23), Greenberg and Mitchell (1983), and Jacoby (1975).

3. Argyris and Schön carefully avoid referring to psychoanalytic issues or unconscious thinking, but their cognitive framework closely parallels the psychoanalytic perspective of this book. Their findings support this book's interpretations. Diamond (1986) offers a thoughtful psychoanalytic critique of the limitations of their views. He points out that people are motivated not by patterns of thinking, but by deeper unconscious, as well as conscious intentions. He emphasizes that people resist changing not simply because they have bad cognitive habits, or think illogically, but because they have unconscious intentions that conflict with their conscious aims. He suggests that individuals are unlikely even to recognize their ineffective thinking patterns until they can acknowledge the unconscious goals they serve.

4. See Schein (1978) for an analysis of matches between workers' career development and organizational roles in psychological contracts. Schein focuses on initial hiring decisions; his study also offers guidance about when a worker may consider leaving.

5. Critics (for example, Klein & Astrachan, 1971) reasonably argue that T groups may simplistically represent interpersonal competence by ignoring important unconscious impediments to collaboration.

6. Possibly, insofar as explicit negotiation reduces some status differences, workers replace parental transferences with unconscious assumptions about brothers and sisters. As a result, dominant feelings of fear of parental power may give way to feelings of sibling rivalry. For an example of fraternal transference, see Hodgson, Levinson, and Zaleznik (1965).

7. Writing in another context, Freire (1970, 1973, 1985) observes that former colonials do not spontaneously act autonomously when formal independence is declared; they have internalized their dependence and found comfort in it. He describes methods through which he has helped these people discover their unconscious domination and learn to accept authority. Although the methods focus on peasants, there are analogies for bureaucratic workers as well.

8. Drawing on Fromm's work, LaBier (1984) argues that a worker's relative health or neurosis may be less important than the fit between the worker's psychological condition and the structure and role expectations of the employing organization. In particular, his research calls attention to two noteworthy groups: "irrational adaptives" and "normal negatives." The former group includes people with personal pathology who, nevertheless, perform effectively and without psychiatric symptoms at work. For example, they may have strong wishes for power over other persons but work in organizations that expect and reward such an interest. The latter group includes people with no special pathology whose work performance is marked by psychiatric symptoms. These people might also work in the organizations just described but lack strong interests in dominating others. LaBier's research offers two optimistic messages with regard to intervention. First, it presents encouraging evidence of apparently satisfying psychological contracts with work organizations. Second, emphasizing the variation in both organizations' psychological structures and workers' psychological orientations, it supports the intervention strategy of attempting to match workers' orientations to organizational structures.

Appendix: The Research

DESIGN OF THE STUDY

The study was designed to probe bureaucratic workers' thoughts and feelings about their organizational work. A relatively small sample of workers were interviewed at length in semistructured interviews. This approach, which draws on the traditional psychiatric interview, is part of the sociological tradition of field research. The semistructured interview has the merit of ensuring coverage of basic issues while permitting respondents the opportunity to express related personal concerns.[1] As noted in Chapter 1, interviews with planners and public administrators focused on different aspects of bureaucratic work. The planners' interviews concentrated on intellectual requirements for problem solving, while the public administrators' interviews concentrated on organizational or political requisites.

Interviews averaged two hours, although some lasted as long as four. The resulting transcripts provide a record of work history, current work issues, and, to varying degrees, connections between personal biography and professional career. Overtly, the transcripts may be interpreted as a discussion of organizational issues, and this content is most amenable to analysis. In addition, the transcripts, as any statements, are implicit records of personal feelings and tacit assumptions about work, relationships, achievement, and satisfaction. Even if these were intentionally clinical interviews, drawing firm interpretations from two to four hours' talk would be difficult. Nevertheless, as the case examples in the book indicate, the interviews are held together by consistent concerns and cohesive cognitive and emotional themes. Interpretation of the transcripts has been informed by the psychoanalytic framework presented in the book.

The interview data are qualitative. Individual responses to questions have been examined for dominant themes. The book presents excerpts from typical statements as the basic unit of the data. Although these interview data do not satisfy traditional canons of quantitative rigor, that is one of the important lessons of the research: much that is important in the bureaucratic experience is ambiguous. Where possible, an effort has been made to suggest the proportions of people who hold particular points of view or feelings. However, both the semistructured character of the interviews and the ambiguity of many issues sometimes make it difficult to refer to any group more specifically than by "most," "many," or "some." Although this imprecision is unsatisfying, the data do not permit greater specificity, and these general qualifiers suggest at least the magnitude of themes among responses.

The research emphasizes intensive analysis of the statements of a limited sample as an appropriate method for exploring the depth and dynamics of the thinking and feeling underlying responses to bureaucratic organizations. Although analysis of a small number of cases inevitably raises questions about the generalizability of findings, the validity of interpretations may be tested in three ways. First, one may examine the consistency of the interpreted responses within the book. Additionally, one may consider the plausibility of the interpretations of these responses in terms of the structural characteristics of bureaucracy. Above all, one may test these interpretations for their usefulness in making sense of others' experiences in bureaucratic organizations.

THE STUDY SAMPLE

The study involves samples from two groups of bureaucratic workers, including 50 city planners and 27 public administrators. The planners were randomly drawn from the membership of the Maryland chapter of the American Institute of Planners, the professional association of planners (now the American Planning Association). The public administrators were randomly drawn from the membership of the Maryland chapter of the American Society for Public Administration, a professional association of workers in public sector organizations. Both organizations comprise primarily workers in public bureaucracies, although both include academics (who also work in bureaucracies), and the planning association includes private consultants. Because of the diffuseness of professional definitions it is difficult to assess the degree to which the memberships in these associations are representative of planners or public administrators, respectively. There is evidence that the composition of the planning association resembles the body of planners enumerated in the national census (American Institute of Planners, 1974, n.d.; Beauregard, 1976). Because of the size of the study samples, it is difficult to assess their statistical representativeness of any larger group, although the planning sample does resemble planners nationally (Baum, 1983). There are no obvious sampling biases that would suggest that these practitioners are significantly different from others who work in public bureaucratic organizations.

Basic characteristics of the samples are summarized in Table A 1. Although there are differences between the two groups, there are important similarities. Because the sample size does not permit comparisons of many differences, it is useful to examine what is common among the practitioners, as an indication of whom else they may represent.

Both groups are predominantly male and white. Both are well-educated, with a vast majority of both holding at least one advanced degree. Most workers are at least in their late thirties, a time when initial commitments to employment have been worked out. Most have been in their present organization long enough to become familiar with its structure and operations. The shorter tenure of the planners is less likely simply a result of their relative youth than a reflection of planners' generally high mobility, which apparently is stimulated by limited opportunities for advancement within the shallow hierarchies of planning agencies. The main differences between the two groups concern the organizations in which they are employed. Planners work in both the public and private sectors; in the public sector they are concentrated at local levels of government. In contrast, the public administrators are relatively evenly distributed among the three levels of government, with a significantly greater representation than planners at the federal level. These differences are reflected in the higher salaries of the public administrators, as well as their accountability to more levels of hierarchy. In this respect, the public administrators work in more typically bureaucratic organizations than do the planners. However, planners' frequent comments about bureaucracy indicate that their organizations may be similarly regarded as bureaucratic.

LIMITATIONS OF THE STUDY

This study of bureaucratic work clearly is exploratory. It examines a small sample intensively. The strength of this approach is that it permits the uncovering of many issues, for which, however, definitive resolution is not possible. As an investigation of the psychological struc-

Table A1. Characteristics of Practitioners Interviewed

	Planners (N = 50)	Public Administrators (N = 27)
Sex		
Male	88%	78%
Female	12%	22%
Race		
White	94%	81%
Black, other	6%	19%
Education		
No college	—	4%
AA only	—	4%
BA only	4%	19%
Master of City Planning or equivalent	74%	7%
Master of Public Administration or equivalent	4%	41%
Other masters degree	26%[a]	33%[a]
JD	6%	4%
Other doctorate	2%	11%
Age		
Range	26–78	26–67
Median	37	43
Years in Organization		
Range	1–20	1–39
Median	5	8
Place of Employment		
Local government	46%	33%
State government	16%	26%
Federal government	6%	37%
Private organization	32%	4%
Levels of Superiors in Organization		
Range	0–5	0–6
Median	2	3
Income		
Median	$20,000–$24,999	$35,000–$39,999

[a]Percentages add to more than 100 because some people earned more than one masters degree

ture of bureaucracy, the study is limited in two ways. One concerns the size and structure of the sample; the other concerns the method of investigation.

First, most of the practitioners studied here work in public bureaucracies. Although it is likely that differences within the public and private sectors are greater than those between the two sectors, the study here does not consider either differences between the structures and tasks of public and private organizations[2] or differences between the people who work in public and private organizations. In addition, the sample size does not permit a number of comparisons of important differences even among the workers interviewed. Two types of variation have received insufficient attention. One is variations among individuals; the other is variations among the organizations in which they work.

In response, another study might both take a larger sample and select respondents differently. This study has drawn samples from groups of practitioners and has then made inferences about organizations from individuals' statements. Because of the variety of organizations it would be useful to attempt to conceptualize differences among organizations and to sample workers according to these categories. Case studies of workers in individual organizations would be consistent with this approach.

The second limitation is that the phenomena studied here—relations between organizational psychological structures and problem solving—are too complex to be encompassed by any single methodology. The types of interviews conducted here capture a great deal of information about individuals, and sensitive interpretation may lead to considerable insight about unconscious concerns. Other questions in similar interviews may accomplish more. Still, more direct investigation of unconscious thinking is required by a variety of methodologies, including free association and projective techniques.[3] At the same time, it should be clear that asking workers to describe their performance and experiences has shortcomings. Workers must also be observed so that their statements may be compared with others' observations.

If the findings from this study are provocative, then these limitations set forth directions for further research.

NOTES

1. For a discussion of the sociological method of field research, see Bogdan and Taylor (1975), Filstead (1970), Glaser and Strauss (1967), Lofland (1971), Schatzman and Strauss (1973), and Zito (1975). For a discussion of the psychiatric interview, see Sullivan (1954; rep. 1970).

2. Rainey, Backoff, and Levine (1976) comprehensively review studies of public and private organizations and tentatively conclude that differences are limited.

3. Maccoby (1976) provides examples of such investigations using several common psychiatric assessment methods. Useful approaches are suggested also by Argyris and Schön (1974, 1978), Levinson (1972), Kets de Vries and Miller (1984), and Mitroff (1983).

References

Altshuler, Alan A. *The City Planning Process; A Political Analysis*. Ithaca: Cornell University Press, 1965.

American Institute of Planners. "AIP membership has not changed much since 1965; average about the same but salaries and education higher." *AIP Newsletter* 9 (1974): 1–3.

———. "Membership Survey from 1976 Roster." Mimeographed. Washington, D.C.: American Institute of Planners, n.d.

Archibald, Kathleen A. "Alternative Orientations to Social Science Utilization." *Social Science Information* 9 (1970): 7–34.

Arendt, Hannah. *The Human Condition*. Chicago: University of Chicago Press, 1958.

Argyris, Chris. *Understanding Organizational Behavior*. Homewood: Dorsey Press, 1960.

———. *Reasoning, Learning, and Action: Individual and Organizational*. San Francisco: Jossey-Bass, 1982.

———, and Schön, Donald A. *Theory in Practice: Increasing Professional Effectiveness*. San Francisco: Jossey-Bass, 1974.

———. *Organizational Learning: A Theory of Action Perspective*. Reading: Addison-Wesley, 1978.

Astrachan, Boris M. "The Tavistock Model of Laboratory Learning." In *The Laboratory Method of Changing and Learning; Theory and Application*, edited by Kenneth D. Benne, Leland P. Bradford, Jack R. Gibb, and Ronald O. Lippitt. Palo Alto: Science and Behavior Books, 1975.

Bacharach, Samuel B., and Lawler, Edward J. *Power and Politics in Organizations*. San Francisco: Jossey-Bass, 1980.

Bales, Robert F. "The Equilibrium Problem in Small Groups." In *Working Papers in the Theory of Action*, by Talcott Parsons, Robert F. Bales, and Edward A. Shils. New York: The Free Press, 1953.

———, and Slater, Philip E. "Role Differentiation in Small Decision-making Groups." In *Family, Socialization and Interaction Processes*, edited by Talcott Parsons and Robert F. Bales. Glencoe: The Free Press, 1955.

Balint, Michael. "Method and Technique in the Teaching of Medical Psychology; II. Training General Practitioners in Psychotherapy." *British Journal of Medical Psychology*, 27 (1954): 37–41.

———. *The Doctor, The Patient, and His Illness*. New York: International Universities Press, 1957.

Banfield, Edward C. "Note on Conceptual Scheme." In *Politics, Planning and the Public Interest*, by Martin Meyerson and Edward C. Banfield. New York: Free Press of Glencoe, 1955.

Bateson, Gregory. *Steps to an Ecology of Mind*. New York: Ballantine Books, 1972.

Baum, Howell S. "Toward a Post-Industrial Planning Theory." *Policy Sciences* 8 (1977): 401–421.

———. "Sensitizing Planners to Organization." In *Urban and Regional Planning in an Age of Austerity*, edited by Pierre Clavel, John F. Forester, and William W. Goldsmith. New York: Pergamon Press, 1980.

———. "Services Aren't Goods: Post-Industrial Principles for Policy Design." *Journal of Sociology and Social Welfare* 8 (1981): 489–512.

———. "Policy Analysis: Special Cognitive Style Needed." *Administration and Society* 14 (1982): 213–236.

———. *Planners and Public Expectations*. Cambridge, Mass.: Schenkman Publishing Company, 1983a.

———. "Politics and Ambivalence in Planners' Practice." *Journal of Planning Education and Research* 3 (1983b): 13–22.

———. "Politics in Planners' Practice." In *Strategic Perspectives on Planning Practice*, edited by Barry Checkoway. Lexington: Lexington Books, 1986.

Beauregard, Robert A. "The Occupation of Planning: A View from the Census." *Journal of the American Institute of Planners* 42 (1976): 187–192.

Bell, Daniel. *The Coming of Post-Industrial Society*. New York: Basic Books, 1973.

Benne, Kenneth D. "Educational Field Experience as the Negotiation of Different Cognitive Worlds." In *The Planning of Change*, edited by Warren G. Bennis, Kenneth D. Benne, Robert Chin, Kenneth E. Corey, 3rd ed. New York: Holt, Rinehart, and Winston, 1976.

———; Bradford, Leland P.; Gibb, Jack R.; and Lippitt, Ronald O., editors, *The Laboratory Method of Changing and Learning; Theory and Application*. Palo Alto: Science and Behavior Books, 1975.

Bennis, Warren G. *Changing Organizations*. New York: McGraw-Hill, 1966.

———; Benne, Kenneth D.; Chin, Robert; and Corey, Kenneth E., editors, *The Planning of Change*, 3rd ed. New York: Holt, Rinehart, and Winston, 1976.

Bennis, Warren G., and Slater, Philip E. *The Temporary Society*. New York: Harper & Row, 1968.

Benveniste, Guy. *The Politics of Expertise*, 2nd ed. San Francisco: Boyd and Fraser Publishing Co., 1977.

Berger, Peter L. *Invitation to Sociology: A Humanistic Perspective*. Garden City: Anchor Books, 1963.

———, and Luckmann, Thomas. *The Social Construction of Reality*. Garden City: Anchor Books, 1967.

Bettelheim, Bruno. *The Uses of Enchantment*. New York: Alfred A. Knopf, 1977.

Bexton, W. Harold. "The Architect and Planner: Change Agent or Scapegoat?" In *Group Relations Reader*, edited by Arthur D. Colman and W. Harold Bexton. Sausalito: GREX, 1975.

Biller, Robert P. "Converting Knowledge into Action: Toward a Post-industrial Society." In *Tomorrow's Organizations*, edited by Jong S. Jun and William B. Storm. Glenview: Scott, Foresman, and Co., 1973.

Bion, Wilfrid R. *Experiences in Groups*. New York: Basic Books, 1961.

Blau, Peter M., and Scott, Richard. *Formal Organizations: A Comparative Approach*. San Francisco: Chandler, 1962.

Bogdan, Robert, and Taylor, Steven J. *Introduction to Qualitative Research Methods*. New York: John Wiley and Sons, 1975.

Bonazzi, Giuseppe. "Scapegoating in Complex Organizations: The Results of a Comparative Study of Symbolic Blame-Giving in Italian and French Public Administration." *Organization Studies* 4 (1983): 1–18.

Bowlby, John. *Attachment and Loss. Volume II. Separation; Anxiety and Anger*. New York: Basic Books, 1973.

Bradford, Leland P.; Gibb, J. R.; and Benne, Kenneth D., editors. *T-Group Theory and Laboratory Method*. New York: John Wiley and Sons, 1964.

Brenner, Charles. *Psychoanalytic Technique and Psychic Conflict*. New York: International Universities Press, 1976.

Burke, Kenneth. *A Grammar of Motives*. Berkeley: University of California Press, 1969.

Burrell, Gibson, and Morgan, Gareth. *Sociological Paradigms and Organizational Analysis*. London: Heinemann, 1979.

Carroll, Lewis. "Through the Looking-Glass." In *The Lewis Carroll Book*, edited by Richard Herrick. 1871. New York: The Dial Press, 1931.

Chassequet-Smirgel, Janine. *The Ego Ideal: A Psychoanalytic Essay on the Malady of the Ideal*, translated by Paul Barrows. New York: W. W. Norton and Co., 1985.

Checkoway, Barry, editor. *Citizens and Health Care: Participation and Planning for Social Change*. New York: Pergamon Press, 1981.

Cherniss, Cary. *Staff Burnout*. Beverly Hills: Sage Publications, 1980.

Churchman, C. W., and Schainblatt, A. H. "The Researcher and the Manager: A Dialectic of Implementation." *Management Science* 11 (1965): B-69–B-87.

Clavel, Pierre. *The Progressive City: Planning and Participation, 1969–1984*. New Brunswick: Rutgers University Press, 1986.

———; Forester, John F.; and Goldsmith, William W., editors. *Urban and Regional Planning in an Age of Austerity*. New York: Pergamon Press, 1980.

Cohen, Arthur H. "Situational Structure, Self-Esteem, and Threat-Oriented Reactions to Power." In *Studies in Social Power*, edited by Dorwin Cartwright. Ann Arbor: University of Michigan, Institute for Social Research, 1959.

Cohen, Mabel Blake, and Cohen, Robert A. "Personality as a Factor in Administrative Decisions." *Psychiatry* 14 (1951): 47–53.

Cohen, Michael D., and March, James G. *Leadership and Ambiguity*. New York: McGraw-Hill, 1974.

Cole, David Bras. *The Role of Psychological Belief Systems in Urban Planning*. Ph.D. dissertation, University of Colorado. Ann Arbor: University Microfilms International, 1975.

Colman, Arthur D., and Bexton, W. Harold, editors. *Group Relations Reader*. Sausalito: GREX, 1975.

Colman, Arthur D., and Geller, Marvin H., editors. *Group Relations Reader 2*, Washington, D.C.: A. K. Rice Institute, 1985.

Crozier, Michael. *The Bureaucratic Phenomenon*. Chicago: University of Chicago Press, 1964.

Cyert, Richard M., and March, James G. *A Behavioral Theory of the Firm*. Englewood Cliffs: Prentice-Hall, 1963.

Dahl, Robert A. *Who Governs? Democracy and Power in an American City*. New Haven: Yale University Press, 1961.

Davies, A. F. *Skills, Outlooks and Passions: A Psychoanalytic Contribution to the Study of Politics*. Cambridge: Cambridge University Press, 1980.

Deal, Terrence E., and Kennedy, Allan A. *Corporate Cultures; The Rites and Rituals of Corporate Life*. Reading: Addison-Wesley, 1982.

Delbecq, Andre L.; Van de Ven, Andrew H.; and Gustafson, David H. *Group Techniques for Program Planning; a guide to nominal group and delphi processes*. Glenview: Scott, Foresman, and Co., 1975.

Denhardt, Robert B. *In The Shadow of Organization*. Lawrence: Regents Press of Kansas, 1981.

———. *Theories of Public Organization*. Monterey: Brooks/Cole Publishing Co., 1984.

Diamond, Michael. "Bureaucracy as Externalized Self-System: A View from the Psychological Interior." *Administration and Society* 16 (1984): 195–214.

———. "Resistance to Change: A Psychoanalytic Critique of Argyris and Schon's Contributions to Organization Theory and Intervention." *Journal of Management Studies* 23 (1986): 543–562.

Dickey, Barbara, and Hampton, Eber. "Effective Problem-Solving for Evaluative Utilization." *Knowledge: Creation, Diffusion, Utilization* 2 (1981): 361–374.

Doktor, Robert H., and Hamilton, William F. "Cognitive Style and the Acceptance of Management Science Recommendations." *Management Science* 19 (1973): 884–894.

Downs, Anthony. *Inside Bureaucracy*. Boston: Little, Brown, and Co., 1967.

Eagle, Jeffrey, and Newton, Peter M. "Scapegoating in Small Groups: An Organizational Approach." *Human Relations* 34 (1981): 283–301.

Edelman, Murray. *Political Language: Words that Succeed and Policies that Fail*. New York: Academic Press, 1977.

Erikson, Erik H. *Childhood and Society*. New York: W. W. Norton and Co., 1963.

———. *Toys and Reasons; Stages in the Ritualization of Experience*. New York: W. W. Norton and Co., 1977.

Etzioni, Amitai. *The Active Society*. New York: Free Press, 1968.

Fenichel, Otto. *The Psychoanalytic Theory of Neurosis*. New York: W. W. Norton and Co., 1945.

Ferenczi, Sandor. "Stages in the Development of the Sense of Reality." In *First Contributions to Psycho-Analysis*, translated by Ernest Jones. 1916. New York: Brunner/Mazel, 1980.

Fiedler, Fred E. "Engineer the Job to Fit the Manager." *Harvard Business Review* 43 (1965): 115–123.

Filstead, William J. *Qualitative Methodology*. Chicago: Markham Publishing Co., 1970.

Fischer, Frank, and Sirianni, Carmen, editors. *Critical Studies in Organization and Bureaucracy*. Philadelphia: Temple University Press, 1984.

Fisher, Roger, and Ury, William. *Getting to Yes; Negotiating Agreement Without Giving In*. New York: Penguin Books, 1983.

Forester, John. "Questioning and Shaping Attention: Toward a Critical Theory of Planning." *Administration and Society* 13 (1981): 161–205.

———. "Planning in the Face of Power." *Journal of the American Planning Association* 48 (1982): 67–80.

———. "Bounded Rationality and the Politics of Muddling Through." *Public Administration Review* 44 (1984): 23–31.

Foucault, Michael. *Discipline and Punish*, translated by Alan Sheridan. New York: Vintage Books, 1979.

Frazer, James George. *The Golden Bough*, abr. ed. New York: The Macmillan Co., 1940.

Freidson, Eliot. *Profession of Medicine*. New York: Dodd, Mead, and Co., 1970.

Freire, Paulo. *Pedagogy of the Oppressed*, translated by Myra Bergman Ramos. New York: Seabury Press, 1970.

———. *Education for Critical Consciousness*. New York: Seabury Press, 1973.

———. *The Politics of Education; Culture, Power, and Liberation*, translated by Donaldo Macedo. South Hadley, Mass.: Bergin and Garvey Publishers, 1985.

French, John R. P., Jr., and Raven, Bertram. "The Bases of Social Power." In *Studies in Social Power*, edited by Dorwin Cartwright. Ann Arbor: University of Michigan, Institute for Social Research, 1959.

Freud, Anna. *The Ego and the Mechanisms of Defense*, translated by Cecil Baines. New York: International Universities Press, 1946.

Freud, Sigmund. *Totem and Taboo*. 1913. Translated by James Strachey. New York: W. W. Norton and Co., 1950.

———. *Introductory Lectures on Psychoanalysis*. 1920. Translated and edited by James Strachey. New York: W. W. Norton and Co., 1977.

———. *Three Essays on the Theory of Sexuality*. 1920. Translated and revised by James Strachey. New York: Basic Books, 1975.

———. *Group Psychology and the Analysis of the Ego*. 1921. Translated and edited by James Strachey. New York: W. W. Norton and Co., 1959.

———. *The Ego and the Id.* 1923. Translated by Joan Riviere. Edited by James Strachey. New York: W. W. Norton and Co., 1962.

———. *Beyond the Pleasure Principle.* 1928. Translated by James Strachey. New York: Bantam Books, 1959.

———. *The Interpretation of Dreams.* 1930. Translated and edited by James Strachey. New York: Avon Library, 1965.

———. *The Problem of Anxiety.* 1936. Translated by Henry Alden Bunker. New York: Psychoanalytic Quarterly Press and W. W. Norton and Co., 1963.

Friedmann, George. *The Anatomy of Work.* Translated by Wyatt Rawson. New York: Free Press, 1964.

Fromm, Erich. *The Sane Society.* Greenwich, Conn.: Fawcett Publications, 1955.

Galbraith, Jay R. *Organizational Design.* Reading: Addison-Wesley, 1977.

Geertz, Clifford. *The Interpretation of Cultures.* New York: Basic Books, 1973.

Getzels, Judith, and Longhini, Gregory. *Planners' Salaries and Employment Trends, 1981.* Planning Advisory Service, Report No. 366. Chicago: American Planning Association, 1982.

Girard, René. *Violence and the Sacred.* Translated by Patrick Gregory. Baltimore: The Johns Hopkins Press, 1977.

Glaser, Barney G., and Strauss, Anselm L. *The Discovery of Grounded Theory: Strategies for Qualitative Research.* Chicago: Aldine-Atherton, 1967.

Glass, James M. *Delusion: Internal Dimensions of Political Life.* Chicago: University of Chicago Press, 1985.

Goffman, Erving. *The Presentation of Self in Everyday Life.* Garden City: Anchor Books, 1959.

Gouldner, Alvin W. *Patterns of Industrial Bureaucracy.* New York: Free Press, 1964.

Greenberg, Jay R., and Mitchell, Stephen A. *Object Relations in Psychoanalytic Theory.* Cambridge, Mass.: Harvard University Press, 1983.

Greenson, Ralph R. *The Technique and Practice of Psychoanalysis.* Vol. 1. New York: International Universities Press, 1967.

Gustafson, James P., and Cooper, Lowell. "Unconscious Planning in Small Groups." *Human Relations* 32 (1979): 1039–1064.

———. "Collaboration in Small Groups: Theory and Technique for the Study of Small Group Processes. In *Group Relations Reader 2*, edited by Arthur D. Colman and Marvin H. Geller. Washington: A. K. Rice Institute, 1985.

Hall, Richard H. *Occupations and the Social Structure.* 2nd ed. Englewood Cliffs: Prentice-Hall, 1975.

———. *Organizations: Structure and Process.* 2nd ed. Englewood Cliffs: Prentice-Hall, 1977.

Harmon, Michael M., and Mayer, Richard T. *Organization Theory for Public Administration.* Boston: Little, Brown, and Co., 1986.

Heilbroner, Robert L. "Economic Problems of a 'Postindustrial' Society." *Dissent* 20 (1973): 163–176.

Hemmens, George C.; Bergman, Edward M.; and Moroney, Robert M. "The Practitioner's View of Social Planning." *Journal of the American Institute of Planners* 44 (1978): 181–192.

Henry, William E. "The Business Executive: The Psycho-dynamics of a Social Role." *American Journal of Sociology* 54 (1949): 286–291.

Hertz, J. H. *The Pentateuch and Haftorahs.* Translated and edited by J. H. Hertz. London: Soncino Press, 1958.

Herzberg, Frederick; Mausner, Bernard; and Snyderman, Barbara. *The Motivation to Work.* New York: John Wiley and Sons, 1959.

Hirschhorn, Larry. "Toward a Political Economy of the Service Society." Working Paper No. 229. Berkeley: Institute of Urban and Regional Development, University of California, 1975.

——, and Krantz, James. "Unconscious Planning in a Natural Work Group: A Case Study in Process Consultation." *Human Relations* 35 (1982): 805–844.

Hodgson, Richard C.; Levinson, Daniel J.; and Zaleznik, Abraham. *The Executive Role Constellation*. Boston: Harvard University, The Graduate School of Business Administration, 1965.

Holland, John L. *The Psychology of Vocational Choice*. Waltham: Blaisdell, 1966.

——. *Making Vocational Choices*. Englewood Cliffs: Prentice-Hall, 1973.

——. *Manual for the Vocational Preference Inventory*. Palo Alto: Consulting Psychologists Press, 1978.

Horney, Karen. *The Neurotic Personality of Our Time*. New York: W. W. Norton and Co., 1973.

Howe, Elizabeth. "Planners' Views of the Public Interest." Paper read at the Annual Meeting of the Association of Collegiate Schools of Planning, 22 October 1983, San Francisco, Cal.: Mimeographed.

——, and Kaufman, Jerome. "The Ethics of Contemporary American Planners." *Journal of the American Planning Association* 45 (1979): 243–255.

Hughes, Everett C. *Men and Their Work*. Glencoe: Free Press, 1958.

Hummel, Ralph P. *The Bureaucratic Experience*. 2nd ed. New York: St. Martin's Press, 1982.

Huysmans, Jan J. B. M. "The Effectiveness of the Cognitive-Style Constraint in Implementing Operations Research Proposals." *Management Sciences* 17 (1979): 92–104.

Jacobi, Jolande. *The Psychology of C. G. Jung; An Introduction with Illustrations*, translated by Ralph Manheim. 8th ed. New Haven: Yale University Press, 1973.

Jacoby, Russell. *Social Amnesia*. Boston: Beacon Press, 1975.

Jaques, Elliott. *The Changing Culture of a Factory*. London: Tavistock Publications, 1951.

——. "Social Systems as Defense Against Persecutory and Depressive Anxiety." In *New Directions in Psycho-Analysis: The Significance of Infant Conflict in the Pattern of Adult Behavior*, edited by Melanie Klein, Paula Heimann, and R. E. Money-Kyrle. London: Tavistock Publications, 1955.

Kafka, Franz. *The Castle*, translated by Willa and Edwin Muir. 1930. New York: Vintage Books, 1974.

——. *Letter to His Father*, translated by Ernest Kaiser and Eithne Wilkins. 1919. New York: Schocken Books, 1966.

Kafka, John. "Challenge and Confirmation in Ritual Action." *Psychiatry* 46 (1983): 31–39.

Kahn, Robert L.; Wolfe, Donald M.; Quinn, Robert P.; Snoek, J. Dedrick; and Rosenthal, Robert A. *Organizational Stress: Studies in Role Conflict and Ambiguity*. New York: John Wiley and Sons, 1964.

Katz, Daniel, and Kahn, Robert L. *The Social Psychology of Organizations*. 2nd ed. New York: John Wiley and Sons, 1978.

Kaufman, Herbert. *The Limits of Organizational Change*. University, Alabama: University of Alabama Press, 1971.

——. *The Administrative Behavior of Federal Bureau Chiefs*. Washington, D.C.: The Brookings Institution, 1981.

Kernberg, Otto. "Notes on Countertransference." *Journal of the American Psychoanalytic Association* 13 (1965): 38–56.

——. *Borderline Conditions and Pathological Narcissism*. New York: Jason Aronson, 1975.

Kets de Vries, Manfred F. R., and Miller, Danny. *The Neurotic Organization*. San Francisco: Jossey-Bass, 1984.

Kipnis, David. "The Powerholder." In *Perspectives on Social Power*, edited by James T. Tedeschi. Chicago: Aldine, 1974.

Klein, Edward B., and Astrachan, Boris M. "Learning in Groups: A Comparison of Study Groups and T Groups." *Journal of Applied Behavioral Science* 7 (1971): 659–683.

Klein, Melanie. "A Contribution to the Theory of Anxiety and Guilt." *International Journal of Psychoanalysis* 29 (1948): 114–123.

———. "Our Adult World and Its Roots in Infancy." *Human Relations* 12 (1959): 291–303.

———. *The Psycho-Analysis of Children*, translated by Alix Strachey. Revised by Alix Strachey and H. A. Thorner. New York: Dell Publishing Co., 1975.

Kohlberg, Lawrence, and Gilligan, Carol. "The Adolescent as a Philosopher: The Discovery of the Self in a Post-Conventional World." In *Twelve to Sixteen: Early Adolescence*, edited by Jerome Kagan and Robert Coles. New York: W. W. Norton and Co., 1972.

———, and Kramer, R. "Continuities and Discontinuities in Childhood and Adult Moral Development." *Human Development* 12 (1969): 93–120.

Kohut, Heinz. *The Analysis of the Self; A Systematic Approach to the Psychoanalytic Treatment of Narcissistic Personality Disorders.* The Psychoanalytic Study of the Child Monographs, Number 4. New York: International Universities Press, 1971.

Kramer, Marlene. *Reality Shock.* St. Louis: C. V. Mosby, 1974.

Krumholz, Norman. "Cleveland: Problems of Declining Cities." In *Personality, Politics, and Planning: How City Planners Work*, edited by Anthony Catanese and Paul Farmer. Beverly Hills: Sage Publications, 1978.

———. "A Retrospective View of Equity Planning: Cleveland 1969–1979." *Journal of the American Planning Association* 48 (1982): 163–174.

———, and Cogger, Janice. "Urban Transportation Equity in Cleveland." In *Metropolitan Midwest: Policy, Problems, and Prospects for Change*, edited by Barry Checkoway and Carl V. Patton. Urbana: University of Illinois Press, 1985.

———; ———; and Linner, John. "The Cleveland Policy Planning Report." *Journal of the American Institute of Planners* 41 (1975): 298–304.

LaBier, Douglas. "Irrational Behavior in Bureaucracy." In *The Irrational Executive*, edited by Manfred F. R. Kets de Vries. New York: International Universities Press, 1984.

Laing, R. D. *The Politics of the Family and Other Essays.* New York: Vintage Books, 1972.

Larson, Magali Sarfatti. *The Rise of Professionalism.* Berkeley: University of California Press, 1977.

Lasch, Christopher. *The Culture of Narcissism; American Life in an Age of Diminishing Expectations.* New York: W. W. Norton and Co., 1978.

———. *The Minimal Self; Psychic Survival in Troubled Times.* New York: W. W. Norton and Co., 1984.

Lasswell, Harold D. *Psychopathology and Politics.* New York: Viking Press, 1960.

———. *Power and Personality.* New York: Viking Press, 1963.

Levinson, Daniel J.; Darrow, Charlotte N.; Klein, Edward B.; Levinson, Maria H.; and McKee, Braxton. *The Seasons of a Man's Life.* New York: Ballantine Books, 1978.

Levinson, Harry. *Organizational Diagnosis.* Cambridge, Mass.: Harvard University Press, 1972.

———; Price, Charlton R.; Munden, Kenneth J.; Mandl, Harold J.; and Solley, Charles M. *Men, Management, and Mental Health.* Cambridge, Mass.: Harvard University Press, 1962.

Lewis, Helen Block. *Shame and Guilt in Neurosis.* New York: International Universities Press, 1974.

Likert, Rensis. *New Patterns of Management.* New York: McGraw-Hill, 1961.

Lindblom, Charles E. "Still Muddling, Not Yet Through." *Public Administration Review* 39 (1979): 517–526.

Lindgren, Henry Clay. *Leadership, Authority, and Power Sharing.* Malabar: Krieger, 1982.

Lippitt, Gordon, and Lippitt, Ronald. *The Consulting Process in Action.* La Jolla: University Associates, 1978.

Lipsky, Michael. *Street-Level Bureaucracy.* New York: Russell Sage Foundation, 1980.

Lofland, John. *Analyzing Social Settings.* Belmont: Wadsworth Publishing Co., 1971.

Lynd, Helen Merrell. *On Shame and the Search for Identity.* New York: Harcourt, Brace, and World, 1958.

Maccoby, Michael. *The Gamesman.* New York: Bantam Books, 1976.

Mack, Ruth P. *Planning on Uncertainty*. New York: John Wiley and Sons, 1971.

Mahler, Margaret S.; Pine, Fred; and Bergman, Anni. *The Psychological Birth of the Human Infant; Symbiosis and Individuation*. New York: Basic Books, 1975.

Malinowski, Bronislaw. "The Role of Magic and Religion." 1931. In *Reader in Comparative Religion; An Anthropological Approach*, edited by William A. Lessa and Evon Z. Vogt. 2nd ed. New York: Harper & Row, 1965.

Mangham, Iain L., and Overington, Michael A. "Dramatism and the Theatrical Metaphor." In *Beyond Method; Strategies for Social Research*, edited by Gareth Morgan. Beverly Hills: Sage Publications, 1983.

March, James G. "Bounded Rationality, Ambiguity, and Engineering of Choice." *Bell Journal of Economics* 9 (1978): 587–608.

———, and Olsen, Johan P. *Ambiguity and Choice in Organizations*. Bergen, Norway: Universitetsforlaget, 1976.

Marcuse, Herbert. "Repressive Tolerance." In *A Critique of Pure Tolerance*, by Robert Paul Wolff, Barrington Moore, Jr., and Herbert Marcuse. Boston: Beacon Press, 1969.

Martin, Joanne. "Stories and Scripts in Organizational Settings." In *Cognitive Social Psychology*, edited by Albert Hastorf and Alice M. Isen. New York: Elsevier/North-Holland, 1982.

———; Feldman, Martha S.; Hatch, Mary Jo; and Sitkin, Sim B. "The Uniqueness Paradox in Organizational Stories." *Administrative Science Quarterly* 28 (1983): 438–453.

Mason, Richard, and Mitroff, Ian. *Challenging Strategic Planning Assumptions*. New York: John Wiley and Sons, 1981.

May, Rollo. *Power and Innocence; A Search for the Sources of Violence*. New York: Dell Publishing Co., 1972.

Mayo, Elton. *The Human Problems of an Industrial Civilization*. New York: MacMillan, 1933.

McClelland, David C. *Power: The Inner Experience*. New York: Irvington Publishers, 1975.

McGregor, Douglas. *The Human Side of Enterprise*. New York: McGraw-Hill, 1960.

Mechanic, David. "Sources of Power of Lower Participants in Complex Organizations." *Administrative Science Quarterly* 7 (1962): 349–364.

Meltsner, Arnold J. *Policy Analysts in the Bureaucracy*. Berkeley and Los Angeles: University of California Press, 1976.

Memmi, Albert. *Dependence; A Sketch for a Portrait of the Dependent*, translated by Philip A. Facey. Boston: Beacon Press, 1984.

Menzies, Isabel E. P. "A Case-Study in the Functioning of Social Systems as a Defense Against Anxiety." In *Group Relations Reader*, edited by Arthur D. Colman and W. Harold Bexton. Sausalito: GREX, 1975.

Merton, Robert K. "Bureaucratic Structure and Personality." *Social Forces* 17 (1940): 560–568.

Michael, Donald N. *On Learning to Plan—And Planning to Learn*. San Francisco: Jossey-Bass, 1973.

Miles, Matthew B. "On Temporary Systems." In *Innovation in Education*, edited by Matthew B. Miles. New York: Teachers College Press, 1964.

Milgram, Stanley. *Obedience to Authority*. New York: Harper & Row, 1974.

Miller, Eric J., and Rice, A. K. *Systems of Organization*. London: Tavistock Publications, 1969.

Mintzberg, Henry. *The Nature of Managerial Work*. New York: Harper & Row, 1973.

———. "Planning on the Left Side and Managing on the Right." *Harvard Business Review* 54 (1976): 49–58.

———. *The Structuring of Organizations*. Englewood Cliffs: Prentice-Hall, 1979.

———. *Power in and Around Organizations*. Englewood Cliffs: Prentice-Hall, 1983.

Mitroff, Ian I. *Stakeholders of the Organizational Mind*. San Francisco: Jossey-Bass, 1983.

————, and Kilmann, Ralph H. "Stories Managers Tell: A Tool for Organizational Problem Solving." *Management Review* (1975) 18–28.

Moynihan, Daniel P. *Maximum Feasible Misunderstanding.* New York: Free Press, 1970.

Myeroff, Milton. *On Caring.* New York: Harper & Row, 1971.

Needleman, Martin, and Needleman, Carolyn Emerson. *Guerrillas in the Bureaucracy.* New York: John Wiley and Sons, 1974.

Orr, Douglass W. "Transference and Countertransference: A Historical Survey." *Journal of the American Psychoanalytic Association* 11 (1954): 621–670.

Paul, Norman L., and Bloom, Joseph D. "Multiple-Family Therapy: Secrets and Scapegoating in Family Crisis." *International Journal of Group Psychotherapy* 20 (1970): 37–47.

Peller, Lili E. "Libidinal Phases, Ego Development, and Play." *Psychoanaltyic Study of the Child* 9 (1954): 178–198.

Perrow, Charles. *Organizational Analysis: A Sociological View.* Belmont: Brooks/Cole Publishing Co., 1970.

————. Review of *Ambiguity and Choice in Organizations*, by James G. March and Johan P. Olsen. *Contemporary Sociology: A Journal of Reviews* 6 (1977): 294–298.

Pettigrew, Andrew M. "On Studying Organizational Cultures." *Administrative Science Quarterly* 24 (1979): 570–581.

Pfeffer, Jeffrey. *Power in Organizations.* Marshfield: Pitman Publishing Co., 1981.

————, and Salancik, Gerald R. "Organizational Decision-Making as a Political Process: The Case of a University Budget." *Administrative Science Quarterly* 19 (1974): 135–151.

Piers, Gerhart. "Shame and Guilt; A Psychoanalytic Study." In *Shame and Guilt*, by Gerhart Piers and Milton B. Singer. 1953. New York: W. W. Norton and Co., 1971.

Pondy, Louis R.; Frost, Peter J.; Morgan, Gareth; and Dandridge, Thomas C. *Organizational Symbolism.* Monographs in Organizational Behavior and Industrial Relations, Volume 1. Greenwich, Conn.: JAI Press, 1983.

Presthus, Robert. *The Organizational Society.* 2nd ed. New York: St. Martin's Press, 1978.

Racker, Heinrich. *Transference and Countertransference.* New York: International Universities Press, 1968.

Radcliffe-Brown, A. R. "Taboo." 1939. In *Reader in Comparative Religion; An Anthropological Approach*, edited by William A. Lessa and Evon Z. Vogt. 2nd ed. New York: Harper & Row, 1965.

Rainey, Hal G.; Backoff, Robert W.; and Levine, Charles H. "Comparing Public and Private Organizations." *Public Administration Review* 36 (1976): 233–244.

Redlich, Fritz C., and Astrachan, Boris M. "Group Dynamics Training." *American Journal of Psychiatry* 125 (1969): 1501–1507.

Reich, Annie. "On Countertransference." *International Journal of Psychoanalysis* 32 (1951).

Rice, A. K. *Productivity and Social Organization: The Ahmedabad Experiment.* London: Tavistock Publications, 1958.

————. *The Enterprise and its Environment.* London: Tavistock Publications, 1963.

————. *Learning for Leadership.* London: Tavistock Publications, 1965.

Rioch, Margaret J. "Group Relations: Rationale and Technique." *International Journal of Group Psychotherapy* 20 (1970): 340–355.

Riviere, Joan. "Hate, Greed and Aggression." In *Love, Hate, and Reparation*, edited by Melanie Klein and Joan Riviere. 1937. New York: W. W. Norton and Co., 1964.

Robinson, John P.; Athanasiou, Robert B.; and Head, Kendra B., editors. *Measures of Occupational Attitudes and Occupational Characteristics.* Ann Arbor: University of Michigan, Institute for Social Research, 1967.

Roethlisberger, Fritz J., and Dickson, William J. *Management and the Worker.* Cambridge, Mass.: Harvard University Press, 1939.

Roheim, Geza. *The Origin and Function of Culture.* 1943. Garden City: Anchor Books, 1971.

Rostand, Edmond. *Cyrano de Bergerac*, translated by Brian Hooker. 1898. New York: Random House, 1951.

Salancik, Gerald R., and Pfeffer, Jeffrey. "The Bases for Use of Power in Organizational Decision-Making: The Case of a University." *Administrative Science Quarterly* 19 (1974): 453–473.

———. "Who Gets Power—and How They Hold on to it: A Strategic-Contingency Model of Power." *Organizational Dynamics* (1977): 3–21.

Sartre, Jean-Paul. *Being and Nothingness*, translated by Hazel E. Barnes. 1943. New York: Washington Square Press, 1968.

Schafer, Roy. *Aspects of Internalization*. New York: International Universities Press, 1968.

———. *A New Language for Psychoanalysis*. New Haven: Yale University Press, 1976.

Schatzman, Leonard, and Strauss, Anselm L. *Field Research*. Englewood Cliffs: Prentice-Hall, 1973.

Schein, Edgar H. *Organizational Psychology*. Englewood Cliffs: Prentice-Hall, 1965.

———. *Career Dynamics: Matching Individual and Organizational Needs*. Reading: Addison-Wesley, 1978.

———, and Bennis, Warren G. *Personal and Organizational Change Through Group Methods: The Laboratory Approach*. New York: John Wiley and Sons, 1965.

Schneider, Susan C., and Shrivastava, Paul. "Interpreting Strategic Behavior: The Royal Road to Basic Assumptions." Working Paper, Number 28. New York: Pace University, Lubin Schools of Business, 1984.

Schön, Donald A. *Beyond the Stable State*. New York: W. W. Norton and Co., 1971.

———. "Some of what a Planner Knows: A Case Study of Knowing in Practice." *Journal of the American Planning Association* 48 (1982): 351–364.

———. *The Reflective Practitioner; How Professionals Think in Action*. New York: Basic Books, 1983.

———; Cremer, Nancy Sheldon; Osterman, Paul; and Perry, Charles. "Planners in Transition: Report on a Survey of Alumni of M.I.T.'s Department of Urban Studies, 1960–1971." *Journal of the American Institute of Planners* 42, (1976): 193–202.

Seeley, John R. "Social Science: Some Probative Problems." In *The Americanization of the Unconscious*. New York: International Science Press, 1967.

Sennett, Richard. *Authority*. New York: Alfred A. Knopf, 1980.

———, and Cobb, Jonathan. *The Hidden Injuries of Class*. New York: Vintage Books, 1972.

Simon, Herbert A. *Administrative Behavior; A Study of Decision-Making Processes in Administrative Organization*. 3rd ed. New York: The Free Press, 1976.

———. *Reason in Human Affairs*. Stanford: Stanford University Press, 1983.

Slater, Philip E. *Microcosm*. New York: John Wiley and Sons, 1966.

Smircich, Linda. "Concepts of Culture and Organizational Analysis." *Administrative Science Quarterly* 28 (1983): 339–358.

Sullivan, Harry Stack. *The Interpersonal Theory of Psychiatry*. New York: W. W. Norton and Co., 1953.

———. *The Psychiatric Interview*. 1954. New York: W. W. Norton and Co., 1970.

———. *Clinical Studies in Psychiatry*. New York: W. W. Norton and Co., 1956.

Susskind, Lawrence; Bacow, Lawrence; and Wheeler, Michael. *Resolving Environmental Regulatory Disputes*. Cambridge, Mass.: Schenkman Publishing Co., 1983.

———, and Weinstein, Alan. "Towards a Theory of Environmental Dispute Resolution." *Boston College Environmental Affairs Law Review* 9 (1980–81): 311–357.

Tannenbaum, Arnold S.; Kavcic, Bogdan; Rosner, Menachem; Vianello, Mino; and Wieser, Georg. *Hierarchy in Organizations*. San Francisco: Jossey-Bass, 1974.

Taylor, F. Kraupl, and Rey, J. H. "The Scapegoat Motif in Society and its Manifestations in a Therapeutic Group." *International Journal of Psychoanalysis* 34 (1953): 253–264.

Thayer, Frederick C. *An End to Hierarchy and Competition: Administration in the Post-Affluent World.* 2nd ed. New York: Franklin Watts, 1981.

Thompson, James D. *Organizations in Action.* New York: McGraw-Hill, 1967.

Thompson, Victor A. *Modern Organization.* New York: Alfred A. Knopf, 1961.

Toennies, Ferdinand. *Community and Society,* translated by C. P. Loomis. 1887. New York: Harper & Row, 1957.

Toker, Eugene. "The Scapegoat as an Essential Group Phenomenon." *International Journal of Group Psychotherapy* 22 (1972): 320–332.

Torbert, William R. "Pre-Bureaucratic and Post-Bureaucratic Stages of Organization Development." *Interpersonal Development* 5 (1974/75): 1–25.

Tuckman, Bruce W. "Developmental Sequence in Small Groups." *Psychological Bulletin* 63 (1965): 384–399.

Van Gennep, Arnold. *The Rites of Passage.* 1908. Translated by Monika B. Vizedom and Gabrielle L. Caffee. Chicago: University of Chicago Press, 1960.

Van Maanen, John. "Experiencing Organization: Notes on the Meaning of Careers and Socialization." In *Organizational Careers: Some New Perspectives,* edited by John Van Maanen. New York: John Wiley and Sons, 1977.

Vasu, Michael. *Politics and Planning.* Chapel Hill: University of North Carolina Press, 1979.

Vroom, Victor H. *Work and Motivation.* New York: John Wiley and Sons, 1964.

Waelder, Robert. "The Psychoanalytic Theory of Play." *Psychoanalytic Quarterly* 2 (1933): 208–224.

Weber, Max. *The Theory of Social and Economic Organization,* 1921 translated by A. M. Henderson and Talcott Parsons, edited by Talcott Parsons. New York: The Free Press, 1964.

———. *From Max Weber,* edited and translated by Hans H. Gerth and C. Wright Mills. New York: Oxford University Press, 1967.

Weick, Karl E. *The Social Psychology of Organizing.* 2nd ed. Reading: Addison-Wesley, 1979.

———. "Managerial Thought in the Context of Action." In *The Executive Mind,* by Suresh Srivastva and Associates. San Francisco: Jossey-Bass, 1983.

Weiss, Carol H. "Knowledge Creep and Decision Accretion." *Knowledge: Creation, Diffusion, Utilization* 1 (1980): 381–404.

White, Orion F., Jr. "The Dialectical Organization: An Alternative to Bureaucracy." *Public Administration Review* 29 (1969): 32–42.

Wiener, Arthur S. "Education and the Post-Industrial Sensibility." Ph.D. dissertation, Wright Institute, 1973.

Wildavsky, Aaron. *Speaking Truth to Power.* Boston: Little, Brown and Co., 1979.

Winnicott, D. W. "Hate in the Countertransference." In *Collected Papers.* New York: Basic Books, 1958.

———. *The Maturational Processes and the Facilitating Environment: Studies in the Theory of Emotional Development.* New York: International Universities Press, 1965.

———. *Playing and Reality.* New York: Basic Books, 1971.

Woodcock, Mike, and Francis, Dave. *Unblocking Your Organization.* San Diego: University Associates, 1979.

Wrong, Dennis H. *Power, Its Form, Bases, and Uses.* New York: Harper Colophon, 1980.

Zaleznik, Abraham. "The Dynamics of Subordinacy." *Harvard Business Review* 43 (1965): 119–131.

———. "Management of Disappointment." *Harvard Business Review* 45 (1967): 59–70.

———. "Psychodynamic Problems in Organizations." Paper read at the First Cornell Symposium on Psychodynamics of Organizational Behavior and Experience, 2 October 1983, New York, New York. Mimeographed.

———; Kets de Vries, Manfred F. R.; and Howard, John. "Stress Reactions in Organizations: Syndromes, Causes, and Consequences." *Behavioral Science* 22 (1977): 151–162.

Zaltman, Gerald; Duncan, Robert; and Holbek, Jonny. *Innovations and Organizations.* New York: John Wiley and Sons, 1973.

Zito, George V. *Methodology and Meaning.* New York: Praeger Publishers, 1975.

Index